M.J. Seymour, Leeds
August

Pelican Books

European Culture and Overseas Expansion

Carlo M. Cipolla was born at Pavia near Milan in 1922. Graduating from Pavia University in 1944, he proceeded to Paris and London, where he continued his studies from 1945 to 1948. Since 1949 he has lectured at various European and American universities on economic history. At present he is Professor of Economic History at the University of Pavia (Italy) and also at the University of California at Berkeley. He is a member of the Academy of Sciences in Turin.

His publications include: *Studi di storia della moneta* (1948), *Mouvements Monétaires dans l'État de Milan* (1951), *Money, Prices and Civilization* (1956), *Le avventure della lira* (1958), *The Economic History of World Population* (1962), *Guns and Sails in the Early Phase of European Expansion* (1965), *Clocks and Culture* (1967) and *Literacy and Development in the West* (1969).

European Culture and Overseas Expansion

Carlo M. Cipolla

Penguin Books

Penguin Books Ltd, Harmondsworth,
Middlesex, England
Penguin Books Australia Ltd, Ringwood,
Victoria, Australia

Guns and Sails first published by Collins 1965
Clocks and Culture first published by Collins 1967

Published in Pelican Books 1970

Made and printed in Great Britain by
C. Nicholls & Company Ltd
Set in Monotype Times

Sic Transit
gloria Mundi

Contents

Prologue 11

Guns and Sails

Introduction 29

1 The European Scene 35

2 Guns and Sails Overseas 73

3 Epilogue 99

Clocks and Culture

1 The European Masters 113

2 Chinese Mandarins and the self-ringing bell 149

3 Appendix on Escapements 171

Bibliography 175

Index 206

Note to the Pelican edition

This book consists of two essays originally published separately in book form: GUNS AND SAILS *In The Early Phase of European Expansion* 1400–1700 (1965) and CLOCKS AND CULTURE 1300–1700 (1967). Professor Cipolla's principal changes for the combined edition have been to make some changes and revisions in the text; to reduce the extent of the footnotes; to omit the illustrations, Richard Ollard's foreword to CLOCKS AND CULTURE, and the author's epilogue to it; to use the Prologue to CLOCKS AND CULTURE as an introduction to both essays and to omit his own prefaces, which also contained his detailed acknowledgements and thanks. He is of course still most grateful to all those who must now be anonymous.

Prologue

A thousand years ago, most of Western Europe was covered by great forests swarming with wild animals and, according to the imagination of the people of the time, with fairies, dwarfs and evil spirits. Marshes too were numerous and widespread and in the south the malarial mosquito created around them a depressing atmosphere of misery and death. The few towns, located within the geographical limits of the old Roman Empire, looked more like villages than towns proper and elsewhere there were no towns at all but only a handful of small and dirty villages.

People were few in number, small in stature, and lived short lives. Socially they were divided among those who fought and hunted, those who prayed and learned, and those who worked. Those who fought did it often in order to rob. Those who prayed and learned, learned little and prayed much and superstitiously. Those who worked were the great majority and were considered the lowest group of all. Work was a mark of low class and the '*ora et labora*' motto of the Benedictines was not enough to ennoble a fatigue that by everybody was considered incompatible with nobility.

The state of the arts was extremely low. Most of the workers were poor and ignorant peasants. Their main solaces in life were getting drunk once a year at the village feast and taking their pleasure at night with their wives, which incidentally helped to compensate for the exceedingly high proportion of children who died in the first years of life. As for the few craftsmen, some of them moved around from one village to another offering their cheap services; others preferred to stay on a manor and obtain a piece of land in exchange for the services that they provided to the local lord or to the community: they occupied part of their time in tilling their land and the rest in turning out some utensils. The manufactures were generally primitive and crude and people looked with astonished admiration at non-European products that on rare occasion they saw in the hands of those wandering and wicked adventurers – the 'merchants' – who, having sold their souls to the devil for the money, moved around from one centre to the other selling objects both common and strange, grains, silks, spices, slaves, jewellery *et similia.*

This drab picture shows signs of improvement after the middle of the tenth century, and more distinctly so after the middle of the eleventh. Population grew, production developed, and in the long run wealth grew more than population. Such processes are not necessarily linked. Especially in the case of poor economies, the growth of population may actually mean the growth of poverty. However, this did not happen in Europe and why it did not happen we do not really know. Our current textbooks of economic history have always some explanation to offer, but even when such explanations are not on the level of the old tale of the fear of the year 1000, they hardly go much further. There are some generalizations one can agree on. Undoubtedly population growth was not rapid, and bottlenecks in the productive system did not prove formidable: there was much land available for the expansion of the agricultural 'frontier' and there were developments in technology. On the other hand, there was expansion both in trade and manufacturing.

This is not the place to speculate about a problem that would need many volumes to deal with adequately, but we must emphasize one fact that was at the same time cause and effect of the development. As far as we can see, urban population grew much more than the rest. This seems to have happened not because fertility was higher or mortality was lower in the cities than in the countryside but because people moved from the countryside to the towns. Within the boundaries of the old Roman Empire, towns that had barely survived during the previous centuries grew larger and had to build and rebuild new walls around their ever-growing areas. Outside old *Romania*, new towns grew up from existing hamlets, or sprang up from the wilderness itself.

Until the Industrial Revolution, there were always more Europeans in the countryside than in towns, but after the eleventh century the proportion of the urban population grew steadily and indeed came to represent the most dynamic element in the development of Western European civilization.

There are reasons to believe that what brought people to the towns were the growing opportunities available in the urban centres rather than any worsening of the economic situation in the country. In fact, I suspect that things were slowly improving also in the countryside because of technological innovation and the evolution and diffusion of the manorial system. People left the country because they thought that in the towns there were better opportunities for economic and social advancement, and this belief made them intolerant of the slow mobility of the rural world. '*Stadtluft machts frei*' was said in Germany: 'the air of the town makes one a freeman'. In more than one way, this movement resembles – as far as motivations and feelings are concerned – the migration of Europeans to America during the nineteenth century. In both cases there was the same hope of moving to a better world, more open socially and full of economic opportunities.

A large portion of those who moved to the towns were unskilled labourers, often young men who then learned their trades in town. But the general economic progress affected rural areas also, artisans became more numerous in the countryside and, at least in the more advanced regions, some villages actually developed traditions of craftsmanship. Since local demand was not enough to support more than a few skilled workers, many of these craftsmen also moved to the towns and enlarged the ranks of the urban working force.

Optimism prevailed in the towns, and it nourished both a general aspiration toward reform and a genuine desire for mutual cooperation: the first took on a religious tone, according to the prevailing spirit of the age, while the second operated mostly on the socio-political level. Guilds grew, and above the guilds a larger form of association developed – the *Commune*. These associations successfully pressed for juridical recognition from the established imperial royal or feudal authorities and whether, peacefully or otherwise, they obtained independent jurisdictional powers. The victory of these associations of free men over the surrounding feudal world was the real turning point in European history. All that happened afterwards was nothing but the logical consequence of this momentous change.

Feudal lords and landed gentry came – or were forced to come – to live in the towns. The urban centres were filled with a large population of priests, monks and nuns. But fundamentally the towns were the domain of merchants and craftsmen, and as the proportion of the urban population over the total population grew in Western Europe in the course of the Middle Ages and the Renaissance, a progressively larger share of the total European population was represented by merchants and craftsmen.

In setting standards of quality for manufacture and in regulating and institutionalizing the training of apprentices, the urban guilds played a positive role in the development of craftsmanship and in the development of a class of skilled artisans. Their importance in this regard, however, should not be exaggerated. Many villages witnessed the growth of a tradition of craftsmanship without having guilds, and on the other hand it has been abundantly proved that the urban guilds soon developed monopolistic tendencies and aimed to limit rather than to expand the supply of skilled labour.

The guilds played a more important role in giving their members a feeling of political importance, a socio-political bargaining power and a social status that as individuals they most certainly did not enjoy in the feudal world. Of course one should not overestimate the 'democratic' aspect of these developments. Social differentiations rapidly appeared on the urban scene. The merchants soon acquired a predominant position and craftsmen often learned that the merchants were no less ruthless to them than the feudal lords had been to their serfs. Among the craftsmen too, social differentiation developed and was often accompanied by conflict. The landed gentry, whether of rural or of new urban formation, retained much power in towns and scarcely distinguished itself for self-restraint. Nevertheless some basic principles had been established, and when one reads that a scholar like Dante had to join a guild in order to be able to play a political role one can understand what a profound difference separated the new urban environment from the old feudal society.

The emergence of a relatively large urban population where merchants, craftsmen amid the professional classes were both numerous and influential gave to the history of the West a characteristic imprint. Sets of values that had prevailed without opposition in the feudal world did not prosper in the new urban one. Utilitarianism began slowly but progressively to permeate European civilization. The realistic reaction of Giotto to the flat, abstract and dogmatic art of the Byzantines, the optimistic and practical spirit brought by Saint Francis to the stream of monastic tradition, were both born of the new social climate. An important aspect of the change was technological progress. This was not as rapid and as brilliant as recent enthusiastic literature may induce the uncritical reader to believe, but it existed and it was substantial. From the eleventh century to the end of the fifteenth, European technology moved ahead in almost every field – in agriculture as well as in the building industry, in navigation and in shipbuilding, in the textile industry, in metallurgy, in carpentry, in accounting, in finance, in transportation, in the production of energy and in warfare. Any book on the history of technology can easily supply the reader with the details of this progress. Here I want to emphasize that, characteristically enough, in medieval Europe a considerable number of anonymous craftsmen experimented in a more or less crude but determined way with wheels, gears and screws of all kinds and descriptions. In antiquity too there had been individuals such as Archimedes and Heron who had had a taste for this kind of gadgetry, but they were odd, isolated cases and their efforts never went beyond the stage of curious experimentation or limited application; machines never became an essential and important element in the productive system. Byzantium and the Mohammedan Middle East followed the Ancients in maintaining the tradition of artistic technology and in both there were craftsmen who dealt with mechanical contrivances. But the number of such craftsmen was always very tiny, while the structure of demand and the prevailing cultural values generally restricted the application of their skills to the production of extravaganzas such as carousels, pneumatic fountains, automatic flutes, etc., that had to satisfy the mania for theatrical splendour of the Byzantine and Moslem rulers. In medieval Europe, on the contrary, elementary applied mechanics became the concern and the object of application of an increasingly large number of craftsmen who dealt with it not out of mere curiosity but from a wish to put it to productive use. Machines came to play an increasingly important role in the productive process.

The mills are a good case in point. Watermills were known in Asia Minor in the first century B.C. and windmills of the vertical type were known in Persia in the seventh century A.D., but it was in medieval Europe, and especially after the tenth century, that a real boom in mill construction developed. A whole series of ingenious mechanical contrivances were devised by anonymous craftsmen to transform the rotary power derived from water or wind into a number of well-differentiated movements of hammers, presses, drills, millstones, and the like. Europe was soon dotted with an astonishingly large number of mills: small mills and large and elaborate ones, windmills, watermills both ashore and afloat, mills for grinding grain and for pressing olives, for fulling cloth and for making paper, for working metal and for brewing beer. The frequency of the family name 'Miller' and the frequent appearance of millers in old tales are abundant witness to the place that the mill acquired in medieval European society; written documents and archaeological remains amply testify to the all-important role of the mills in the most important sectors of the economy. Clocks and automata offer another good example, and this subject will be discussed extensively in the first chapter of the essay *Clocks and Culture*.

The existence of numerous crowded urban centres with relatively low standards of public health and hygienic conditions made the European population highly vulnerable to epidemic disease. About the middle of the fourteenth century a pandemic of plague devastated Europe and allegedly killed one-third of its inhabitants. Throughout the following decades and centuries, plague and various other kinds of epidemics were a recurrent unremitting tragic theme in European history. As a result the growth of population was checked. We lack precise statistical information but it is probably safe to say that at the end of the fifteenth century the population of Europe was no greater than at the beginning of the fourteenth.

One of the effects of the lack of a substantial population growth in the presence of growing economic activity must have been a substantial increase in real wages. The economic position of the craftsmen improved, and they pressed strongly or even violently for a more active part in the administration of the commonwealth. On the other hand, since labour was becoming more expensive, the pressure to adopt labour-saving devices grew stronger and the Europeans became more machine-minded than ever.

When one tries to describe in a very few pages a set of changes that took place over many centuries and transformed a whole continent, one is bound to be prey to over-simplification and deservedly becomes subject to serious criticism. In the preceding pages I have tried to single out some basic trends that I think were of great importance in shaping modern European civilization. But even if one leaves aside the question of regional differences, to which I shall return later, it is abundantly clear that the previous pages scarcely do justice to all the complexities of the changes considered. Terms such as 'merchants' and 'craftsmen' are exceedingly inadequate to represent a multifarious reality of different human and social characters. It should also be mentioned that the change was painfully slow, especially in its beginnings. Only in the course of time did it acquire momentum. This progressive acceleration was the result of the fact that, as so often happens in social processes, the change was characterized by cumulative mechanisms. For instance, the availability of able craftsmen and the existence of a developed technology in Western Europe explains the rapid proliferation of fire-arms as well as the improvements in shipbuilding and the expansion of ocean navigation after the end of the fourteenth century. In its turn, the rapidly growing demand for cannon, cannon balls and anchors stimulated the growth of both the metal and the mining industries, adding new impetus to the proliferation of skilled artisans and to the progress of technology. The development of mining, on the other hand, stimulated the search for mechanisms that would help human muscle-power or replace it in the pumping of water out of the mines and in the hauling of minerals. Similarly, technological progress and interest in technological matters gave birth to a rich literature devoted to mechanical problems. The invention of printing, itself a result of technological progress, helped the diffusion of technical literature and fostered the pace of technological advance.

It must also be said that economic development progressively enriched and diversified social groups. Among the artisans, a group of superior craftsmen gradually emerged (think of Brunelleschi, Francesco di Giorgio or Leonardo) and separated themselves from whitewashers, stonedressers, masons and smiths. They formed connexions with high circles in society and actively broadcast their interests and views. Interest in the machine grew progressively stronger and was always characterized by those feelings of utilitarianism that prevailed in the medieval urban environment.

> Without mechanical ingenuity the force of man is of small avail: in building it is necessary to move great weights; likewise there is need to draw water in large quantity and convey it over long distances; no less useful and necessary will be the construction of mills, and in some places where little water is available, ingenuity must be used to help; in other places where there is no water at all, mills must be constructed that work with the wind or by some other means.

These words were not written by a professional engineer: there were no professional engineers in the fifteenth century, because specialization had not yet succeeded in putting men into labelled boxes and their brains into strait-jackets, although the trend was taking shape. The words I have quoted were written by Francesco di Giorgio Martini (1439–1501) who, trained as a craftsman, successful as a painter, and expert in architecture, ended up by writing, in accordance with the mood of the age, a treatise on engineering. It was as 'natural' for the people of the Renaissance to bother their heads with perpetual motion machines, mechanical pumps, water scoops, war machines, mills and automata, as it was 'natural' for the Greeks of Plato's day not to bother theirs. One may remember the technical treatises by Fontana (1420), the treatise *De Machinis* by I. Mariano (1438–41), the writings on architecture by Alberti, Filarete and Palladio, the work on military machines by Volturio da Rimini (1472), the *Unterrichtung zur Befestigung der Städte* by Dürer (1527), the *De Tradendis Disciplinis* by Vives (1521), the treatises by Biringuccio on metallurgy (1540), by G. Agricola on mining (1556), by Guidobaldo del Monte on applied mechanics (1577), the *Theatrum Instrumentorum* by J. Besson (1578), the *Diverse e Artificiose Macchine* by Ramelli (1588), the *Pneumaticorum Libri* by Della Porta (1601), the *Teatro Nuovo di Machine et Edificii* by Lonca (1607). If Leonardo da Vinci had not existed, the history of European technology would have been the same, but his drawings and sketches witness in a superb way the extravagant interest in wheels, toothed wheels, gears and the like that took over in Europe.

When in 807 the great Haroun al Rashid sent an embassy to Charlemagne, among the presents brought to the western emperor was 'a clock made with wonderful mechanical skill (*arte mechanica mirifice compositum*) driven by water and showing the twelve hours which are sounded by an appropriate number of small bronze bells dropping into a brass basin. At noon twelve horsemen come out of twelve windows which close behind them.' Eginhard (*Annales*, ad annum) to whom we are grateful for this information added that 'there are many other things in this clock,' and his words disclose the astonishment and the admiration that the Arabian clock aroused at the Frankish court. Nowhere in the West was it possible to make or to see such mechanical wonders. In 949, when Liudprand of Cremona visited Constantinople, he was admitted to the presence of the Greek emperor in a palace of remarkable size and beauty, named Fresh Breeze.

Before the Emperor's seat stood a tree, made of bronze gilded over, whose branches were filled with birds, also made of gilded bronze, which uttered different cries, each according to its varying species. The throne ... was of immense size and was guarded by lions, made either of bronze or of wood covered over with gold, who beat the ground with their tails and gave a dreadful roar with open mouth and quivering tongue. Leaning upon the shoulders of two eunuchs I was brought into the emperor's presence. At my approach the lions began to roar and the birds to cry out. ... After I had three times made obeisance to the emperor with my face upon the ground, I lifted my head, and behold! the man whom just before I had seen sitting on a moderately elevated seat had now changed his raiment and was sitting on the level of the ceiling. How it was done I could not imagine, unless perhaps he was lifted up by some such sort of device as we use for raising the timbers of a wine press (*Antapodosis*, VI, 5).[1]

Again the mechanical skill and ingenuity of the East caused wonder and admiration in an educated Western cleric.

At the beginning of the thirteenth century there was still nothing in Western literature that could compare with Al-Jazari's technological encyclopedia, composed about 1205 by this outstanding craftsman after twenty-five years spent at the court of the Urtuqid Sultan of Diar-Bekr on the Tigris. The Byzantines of the thirteenth century did not find any reasons to modify those ideas to which they had been attached for centuries: namely that the 'Latins' (read Westerners) were just one kind of 'barbarians' and that the 'New Rome' (Constantinople) was so far superior to the Old that to compare the two would be ridiculous. Visiting distant China, Marco Polo felt like a 'barbarian' among more developed people. But the situation was rapidly changing. The sketch book of Villard de Honnecourt gives a significant indication of what was emerging in the West and two centuries later the technological pre-eminence of Western Europe over Asia was an accomplished fact. At the end of the fourteenth century, some Byzantines like Demetrius Cydones, taking a closer look at the Western culture, made the exhilarating discovery that the West had something to offer to the East. Decades later, the Greek-born Cardinal Bessarion addressed to Constantine Palaeologos, despot of the autonomous Byzantine province of Morea, a memoir in which he spoke bluntly of Western superiority. In this memoir Bessarion urged Constantine to send young men to Italy.[2] These young men should 'learn the rudiments of some skills' and bring to Greece the more advanced Western technology especially in the fields of mechanical engineering, shipbuilding, the manufacture of arms and the iron industry. 'Mechanical engineering', wrote the Cardinal,

facilitates the drawing of heavy objects, the demolition of ruins, the grinding of what needs grinding: boards are sawn automatically; mills turn as rapidly and exactly as possible; in metallurgy bellows are in-

1. Wright, *The Works of Liudprand*, pp. 207–8. Further details of books mentioned in the footnotes are given in the Bibliographies at the end of the book.

2. Lambros, *Ipomnina*, p. 26.

flated and deflated without the touch of human hands, separating the metal from the baser material in the flux.

The skill of making iron, so useful, so necessary to mankind, the absence of which allows nothing to go smoothly whether in war or in man's peaceful political life, this too one might easily learn here at Rome. I hear that the Peloponnese, and especially the district around Sparta, abounds in the ore of iron. Yet of all the necessities of life, Sparta lacks iron manufactures and must import them from elsewhere.

The preparation and manufacture of armaments, both offensive and defensive, without which no war can be carried on satisfactorily, you can put into effect for yourself and the rest of Greece, especially if you have the iron in your country and acquire in addition those skilled in its manufacture.

You ought naturally to value highly the art of shipbuilding also, since its benefits are most advantageous, and not only the building of warships and triremes, but also of merchant ships, the timber for which abounds in the Peloponnese and is of the highest quality.

These four skills . . . mechanics, the manufacture of iron, the manufacture of arms, and shipbuilding, necessary and useful as they are for those who wish to live well, you would be able to import into Greece and transfer to our people through the expedient of having four or eight young men introduced here (in Italy) in a suitable and unobtrusive way.

Indeed, there are four other skills worthy of mention: the manufacture of glass, silk-weaving, the manufacture of woollen cloths and in addition the art of dyeing both. But since these are skills which were devised rather for luxury and amusement than for the sheer necessities of men, I lay little stress on these before the acquisition of the skills which are absolutely necessary.

When the letter by Cardinal Bessarion was being written, Italy – or, more accurately, northern and central Italy – was undoubtedly one of the leading European countries in matters of craftsmanship and technology. That a humanist such as Cardinal Bessarion should choose Italy, the cradle of humanism, as the example to offer for the education and development of the young generations of his own country is perfectly justified and completely understandable. Italy, however, was not the only area in Europe to enjoy at that moment a developed technology and a relative abundance of craftsmen. The Low Countries, especially the southern Low Countries, were from this point of view no less developed than Italy. France, too, had plenty of skilled technicians. Her development had been hampered by the destruction and the disorder caused by the Hundred Years War (1337–1453), but she recovered very rapidly and brilliantly in the second half of the fifteenth century. At almost the same time, thanks to the influences and impulses radiating from the Low Countries and from Italy, western and southern Germany experienced a remarkable development; by the end of the fifteenth century the Germans were masters in metallurgy and mining.

The areas just mentioned were the cradles of late medieval and early Renaissance technology. On the fringe, other countries took part in a more or less passive way in the general European development, but by the standards prevailing in the Low Countries, in western and southern Germany, and in northern and central Italy they could easily be defined as underdeveloped. Spain had some nuclei of good craftsmanship in Biscay, Catalonia and Toledo, but Catalonia entered a period of decline after the middle of the fourteenth century. In general, one could say with Guicciardini, at the beginning of the sixteenth century, that poverty prevailed in Spain 'not so much for the quality of the country, but because the Spaniards by nature are not inclined to the arts and crafts'. The Scandinavian countries were no better off. England was in somewhat better shape; she had developed a fairly good textile industry and was the most advanced country on the periphery of Europe. However, she was far behind France, Germany, Italy and the Low Countries. In the first half of the sixteenth century she was still heavily borrowing craftsmen and techniques from the Continent for a number of trades, particularly mining and metal working.

The basic asset of the more developed countries was their stock of human capital, namely their relatively large number of active merchants and good craftsmen. This capital was the result but at the same time also the cause of growth and development, and throughout the Middle Ages and the Renaissance rulers and administrators not only fussed over the imports and exports of foodstuffs and bullion: they were also deeply concerned about the inflow and outflow of craftsmen and the local supply of able artisans.

The mobility of skilled labour before the Industrial Revolution is a topic that has not been adequately investigated by economic historians and we know relatively little about it. There is no doubt, however, that the phenomenon existed on quite a large scale and that it was of great importance in the technological and economic history of Europe. I have already mentioned that skilled artisans often left their villages. Urban craftsmen were no less mobile. We often hear of master-artisans or groups of master-artisans who left one town for another or one country for another, fleeing from political troubles, from an epidemic, or from an economic crisis. '*Vagabundi sunt ut aves*' ('They are vagabonds, like the birds') was said of peasants in a medieval document, but the same thing could be said of craftsmen.

Religious fanaticism and a series of major European wars greatly fostered the mobility of labour in the course of the sixteenth century. Italy, which had attracted numerous foreign craftsmen throughout the previous centuries, was the first to suffer from the political and religious upheavals of the sixteenth century. The Franco-Spanish Wars (1494–1559), with their tragic trail of famines, epidemics and depression, induced many skilled artisans to leave the country. This exodus tapered off somewhat in the second half of the sixteenth century but it resumed strength during the critical decades of the seventeenth century when the country's economy collapsed. France profited from a large influx of Italian craftsmen during the first part of the sixteenth century but after 1560 she was torn apart by feuding religious and political groups and lost many skilled workers. In the southern Low Countries, Spanish conquest and occupation, Catholic fanaticism, and Dutch blockades proved disastrous: the whole economy collapsed in the course of the second half of the sixteenth century and capable craftsmen left the country in increasing numbers. Germany supplied the rest of Europe with skilful craftsmen throughout the fifteenth, sixteenth and seventeenth centuries. It is impossible to give meaningful global figures about these movements and their directions, but we can say with a fair degree of plausibility that Italy, the southern Low Countries, France and Germany were on the whole the areas that lost precious human capital, while the Dutch United Provinces, England, Switzerland and, later, Sweden were the countries that profited correspondingly.

If the stimulus of skilled immigrants was to have lasting effects on an economy, the recipient country had to be open to new ideas and new techniques. Many Italian technicians went to Turkey in the course of the fifteenth and sixteenth centuries yet nothing happened to the Turkish economy. On the other hand, the emigration of skilled artisans, while contributing to the decline of a country, was also a symptom of it. Italy, for instance, which lost many able craftsmen, had turned from a dynamic and highly receptive society into an obtusely conservative one in the course of the sixteenth and seventeenth centuries. The Italians (of the late Renaissance) wrote Fynes Moryson at the beginning of the seventeenth century,

> thincke themselves to have somuch understanding, and their Country to yealde somuch sweetenes, fruitfullness and such monuments of arts and fabricks, as they seldome or never travaile into forayne Kingdomes, but driven by some necessity ether to followe the warrs or to traffique abroade: this opinion that Italy doth afforde what can be seene or knowne in the world, makes them only have homebred wisdome and the prowde conceete of their owne witts . . .[3]

As it happened, the countries which were receiving skilled labour also adopted a new version of the Christian faith whose essential bibliolatry fostered the diffusion of literacy, thus contributing to the qualitative improvement of the stock of human resources. The combination of these and other factors played a decisive role in changing the balance of economic power in Europe between 1550 and 1650. Two centuries after Cardinal Bessarion urged the youth of his country to go to Italy to learn the best technology of the West, leadership in technological and economic development had been taken over by England, Holland, and, as far as metallurgy was concerned, Sweden.

In the meantime, important developments were taking place on the cultural level. The official seats of higher learning – the Universities – had remained relatively untouched by the cultural changes that the prevalence of merchants and craftsmen had brought about. The study of law, of course, was influenced by the development of trade, and theology had to come to terms with

3. Moryson, *Itinerary*, cd. Hughes, p. 419.

some of the needs of a commercialized society, but fundamentally the Universities remained strongly tied in their outlook to a distant past. The majority of medieval scholars scarcely recognized technological progress or showed any interest in it.

But the tide was turning with the rise of experimental science after the end of the Middle Ages. One can easily agree with Dr R. Hall when he denies 'that the behaviour of an empirical scientist is derivable by virtually direct imitation from the trial-and-error, haphazard and fortuitous progress of the craft' and one can also agree with Dr Hall's other statement that 'while scholars showed increasing readiness to make use of the information acquired by craftsmen and their special techniques for criticizing established ideas and exploring phenomena afresh, it is far less clear that craftsmen were apt or equipped to criticize the theories and procedures of science.'[4] But it would be a *non sequitur* to argue that the social and technological environment of the times played no role whatsoever in the intellectual change among the learned. One must be careful to avoid putting things in terms of a naïve dichotomy in which full-grown scientists face simple craftsmen. History works in a more subtle way and through a much more complicated network of channels. Before becoming a full-grown scientist, a man must go through the state of childhood and adolescence when he is influenced by what he sees and hears in his family circle, among his friends and in the street. Dr Hall maintains that the spectacular achievements of applied technology did not 'arrest the attention of scholarly scientists',[5] but all that had been accomplished with gunpowder, the magnetic compass, and the press most certainly influenced the environment in which the young men who were the scientists-to-be grew up and were formed.[6] If people such as Francesco di Giorgio and Leonardo, who were above all and essentially artists, delighted in drawing mills, gears and machines instead of flowers, fishes and butterflies as their contemporary Chinese fellow painters did,

4. Hall, *The Scholar and the Craftsman*, p. 21.

5. ibid., p. 15.

6. In the writings of the sixteenth and seventeenth centuries the printing press, gun-powder and the compass are often quoted as great innovations that changed the course of history. This contradicts the statement by Hall that the achievements of applied technology did not arrest the attention of the scholarly scientist.

the explanation lies in those anonymous but overpowering influences that the environment exerts on men from their childhood on, in shaping their tastes and their values. Similarly, if one pauses to consider the utilitarianism that pervaded the outlook of the new scientists, one can hardly help being reminded of the effects of a social environment in which merchants and craftsmen had been for a long time the most vigorous and dynamic group.

Although much about the origins of the Scientific Revolution remains obscure, we know much about its essence and its main aspects. An increasing number of scholars asked questions within the range of an experimental answer by limiting their inquiries to physical rather than metaphysical problems and by concentrating their attention on accurate observation of the kinds of things that are in the natural world. At the same time they accepted as a goal of science that subjection of the natural environment that has formerly seemed to belong only to art and the crafts. Their success was based on the combined application of measurement, mathematics and experiment, and it was precisely those branches of science which were most amenable to measurement that showed the most spectacular achievements. Resistance to change was not weak or easy to overcome. If there were passionate champions of the new science, there were no less passionate advocates of traditional scholarship. The conservatives clustered quite naturally around established institutions, namely the Universities. The champions of the new science founded new institutions, the Academies, and clustered around them. In the long run, the innovators won. Their victory was the victory of a new philosophy imbued with empiricism and utilitarianism that infected all branches of human knowledge. Mathematics was to be the main tool of analysis and the machine was the reference-idea. The whole 'management of this great machine of the world,' wrote H. Powers in 1664, 'can be explained only by the experimental and mechanical Philosophers.'[7] Indeed, what started then was the 'mechanization of the world view'.

The utilitarian spirit, born of the medieval urban civilization, fostered by the humanism of the Renaissance and narrowed and emphasized by the Baconian philosophy, expressed itself in an

7. Quoted by Jones, *Ancients and Moderns*, p. 204.

ever-growing craze for new machines and in an avid interest in those manual crafts that were responsible for the making of such contraptions. On the other hand, mechanics, chemistry, microscopy, qualitative astronomy were still in their infancy and there were no important barriers blocking entry into these new areas of exploration. Gabriel Harvey wrote in 1593 that any 'expert artisan or any sensible industrious practicioner, howsoever unlectured in schooles or unlettered in bookes' could make notable contributions to the advancement of science. This was a rosy overstatement. About a century later more realistically Evelyn complained in a letter to Boyle of the 'many subjections which I cannot support, of conversing with mechanical capricious persons'.[8] Nevertheless one can quote a number of cases in which ideas and suggestions were exchanged among scientists and highly-skilled craftsmen such as clockmakers, lens-cutters and precision-instrument makers. Professor Needham is essentially correct when, contrasting the European situation with the situation prevailing in China, he writes that: 'In Europe, unlike China, there was some influence at work ... that pushed forward to make the junction between practical knowledge and mathematical formulations.... Part of the story undoubtedly concerns the social changes in Europe which made the association of the gentleman and technician respectable.'[9] In fact a good part of the story concerns also the socio-cultural changes that made it respectable for the gentleman to apply himself to technical and scientific matters. Besides scholars and craftsmen there was in Europe a large and growing group of amateur-scientists, from the middle and upper classes, who were neither professional scholars nor craftsmen. The role of these *virtuosi* in fostering the progress of science during the seventeenth and early eighteenth centuries can hardly be overstated and was certainly greater than that of the artisans.

The Chinese had invented the printing process many centuries before the Europeans discovered it, but it was the Europeans who exploited it to the full. This is in fact a good case to illustrate the point that while an important technological innovation has a

8. Boyle, *Works*, vol. 6, pp. 287–8.
9. Needham, *Science and Civilization in China*, vol. 3, pp. 154–5.

good chance to influence or modify the socio-cultural environment, the ultimate effects of the same innovation rest on the nature and the quality of the environment. By the early seventeenth century, the great majority of the Chinese population was still illiterate while in Europe literacy had made remarkable progress. The large group of amateur-scientists who substantially contributed to the flourishing of Academies and to the progress of the Scientific Revolution would hardly have existed if literacy had been restricted to a few high priests and if printing had not developed into a flourishing industry.

As long as technological progress was mostly in the hands of common craftsmen who were guided mainly by tradition and some rough rules of trial-and-error, the rate of progress could never have been very high, but it accelerated dramatically when the resources of craftsmanship were strengthened by the systematic application of scientific principles developed by more or less professional scientists. It is enough to go through the records of patents issued in England, Holland and France in the course of the seventeenth century to realize how important the change was. New machines, instruments for navigation and for measurement, barometers, thermometers, microscopes, telescopes, contraptions of all kinds and descriptions were invented, and developed. It was in this environment that von Guericke's pneumatic pump – the ancestor of the steam engine – was born. When all these developments are considered, it is not at all surprising that some decades later the Industrial Revolution took place – in fact, it would be highly surprising if it had not.

Guns and Sails

Introduction

On the 28th day of May 1453 the Turks entered Constantinople. All Europe trembled, overwhelmed by amazement and dismay. 'A thing terrible to relate, and to be deplored by all who have in them any spark of humanity, and especially by Christians,' wrote Cardinal Bessarion to Francesco Foscari, Doge of Venice,

> A city which was so flourishing, with such a great empire, such illustrious men, such very famous and ancient families, so prosperous, the head of all Greece, the splendour and glory of the East, the school of the best arts, the refuge of all good things, has been captured, despoiled, ravaged, and completely sacked by the most inhuman barbarians and the most savage enemies of the Christian faith, by the fiercest of wild beasts.... Much danger threatens Italy not to mention other lands, if the violent assaults of the most ferocious barbarians are not checked.

One can easily understand the shock experienced by contemporary Europeans and the wave of fear that the event spread throughout Europe. Yet, there was nothing essentially new. Europe had lived a chronically precarious life throughout the centuries of the Middle Ages always at the mercy of potential invaders. Granted, things were not so bad in the later as they had been in the early Middle Ages. The Moslems had been completely expelled from the Iberian Peninsula as well as from Southern France and Southern Italy. The Vikings and the Hungarians had been assimilated. Large territories had been acquired east of the Elbe river. However, the general balance of power had not turned in favour of Europe, and in the long run the Europeans remained on the defensive.

The Crusades should not mislead us. The success that characterized the first phases of the European attack was largely due to surprise and to the temporary weakness and disorganization of the Arab world. As Grousset once said it was the victory of 'French monarchy over Moslem anarchy', but the Moslem forces were soon reorganized and the Europeans had quickly to retreat. Echoing a speech by Urban II and summing up the experience of the First Crusade, William of Malmesbury wrote: 'This little portion of the world which is ours is pressed upon by warlike Turks and Saracens: for three hundred years they have held Spain and the Balearic Islands and they live in hope of devouring the rest.'

The aggressiveness and success in expansion that the West showed after the eleventh century on the commercial plane, did not have a counterpart in the military and political field. The disaster of Wahlstatt in 1241 dramatically proved that Europe was militarily incapable of coping with the Mongol menace. If Europe was not invaded, it was because of the timely death of the Mongolian chief (Ögödäi, December 1241) and because, in the long run, the Khans were more attracted by the South and the East than by the West. In the following century, the defeat of the Christians at Nicopolis (1396) showed once again the military weakness of the Europeans in face of Eastern invaders. Again Europe was saved by purely fortuitous circumstances: Bāyāzed, the conqueror, got involved with the Mongols of Timur Leuk (Tamerlane) and one potential danger luckily and unexpectedly eliminated the other. In the fifteenth century Europe was still under the pressure of Turkish attack, and although at times capable of slowing down the advance of the enemy, she never succeeded in stopping it altogether.

The reasons for the chronic weakness of medieval Europe are clear enough. To begin with, the Europeans were not very numerous (never more than 100 million people). More important than that, they were divided and constantly busy in 'waging wars against each other, staining their hands with the blood of their own people, defiling their arms with the blood of Christians'. When composite armies were put together the main result was general confusion. Last but not least, the military organization of the European potentates was far from being efficient. Europe and more especially Eastern Europe relied on heavily armoured cavalry which was colourful but unwieldy. As has been said,

'European nobility sacrificed tactics and strategy for the impossible dream of striking heavily at the enemy while remaining invulnerable.' For one reason or another, throughout the Middle Ages, Europe's main hope of survival continued to be placed in large measure in the hands of God.

After the fall of Constantinople things became progressively worse. The Turkish advance continued powerful and seemingly irresistible. Northern Serbia was invaded in 1459. Bosnia-Herzegovina in 1463–6. The Negroponte was taken from the Venetians in 1470. Albania was invaded after 1468. 'I cannot persuade myself that there is anything good in prospect,' wrote Pope Pius II. 'Who will make the English love the French? who will unite the Genoese and the Aragonese? who will reconcile the Germans with the Hungarians and Bohemians? If you lead a small army against the Turks you will easily be overcome; if a large one, it will soon fall into confusion.'

Yet, at the moment when the enemies of Christendom seemed to be striking at the very heart of Europe, a sudden and revolutionary change occurred. Outflanking the Turkish blockade, some European nations launched successive waves of attack over the oceans. Their advance was as rapid as it was unexpected. In little more than a century, the Portuguese and the Spaniards first, the Dutch and the English later, laid the basis of world-wide European predominance.

Some decades ago it was fashionable to argue that the oceanic exploration and expansion of Europe in the second half of the fifteenth century were the direct consequence of the advance of the Turks who had interrupted the flow of spices to Europe via the Near East. This view was an excellent piece of historical naïvety and it actually proved to be totally incorrect, however, as it often happens, mixed with the error, one may possibly find also a grain of truth. One way to look at the European quest for the direct routes to the Spice Islands and the coasts of West Africa is to consider these phenomena as an aspect of the tension between the economic expansion of Europe and the military and political blockade that was imposed upon her.

However, no matter how strong the tension, the motive is one thing and the instrument by which the motive can be transformed into effective and successful action is quite another. The need to outflank the Moslem blockade and reach the Spice Islands was already felt in the thirteenth and fourteenth centuries. But the fact that the Atlantic expeditions of the brothers Vivaldi and of Jaime Ferrer failed, is a proof – if a proof is needed – that although there were 'motives', the necessary 'means' were not available. It has also been written that the Europeans 'went into Asia in a spirit of determination to succeed which was stronger than the will of the Asiatic people to resist' and that this mainly accounted for European success. But no spirit of determination, strong as it may be, is enough to win a battle if the necessary means are not available. The brothers Vivaldi certainly did not lack the 'spirit of determination', but their galley was not fitted to brave the oceans. Why did Renaissance Europe succeed where the Europeans of the thirteenth and fourteenth centuries failed? Why, after the end of the fifteenth century, were the Europeans able not only to force their way through to the distant Spice Islands but also to gain control of all the major sea-routes and to establish overseas empires? What allowed the Europeans to accomplish the dramatic and sudden transition from a state of insecure defence to that of bold and aggressive expansion? Why did the 'Vasco da Gama era' come into being?

Chapter 1

The European Scene

As early as the first decades of the fourteenth century Europeans began to use cannon in warfare. Florentine official documents refer to the acquisition of '*pilas seu palloctas ferreas et canones de mettallo*' in 1326, thus indicating that by that date bronze guns shooting iron balls were already in use. In 1327 an illuminated manuscript was produced in England containing the picture of what is undoubtedly a very primitive gun. Unfortunately the text makes no reference to the illustration, but at the Statens Historiska Museum of Stockholm there is a rare piece which corresponds to the gun of the Millimete manuscript. This is the earliest evidence of European artillery in the modern sense of the word. After 1330 guns were commonly used in warfare and in the 1350s Petrarch (*De Remediis*, lib. 1, dialog. 99) wrote that 'these instruments which discharge balls of metal with most tremendous noise and flashes of fire . . . were a few years ago very rare and were viewed with greatest astonishment and admiration, but now they are become as common and familiar as any other kind of arms. So quick and ingenious are the minds of men in learning the most pernicious arts.'

In the second half of the fourteenth century, normal cannon was well established in warfare, and European metal workers were striving for super-weapons. Huge guns of the bombard type were then developed and became very popular. As has been written, 'the most striking occurrence in the early history of cannon is perhaps the great size which they soon attained'.[1] In 1382 at the siege of Oudenarde the army of Philip van Artevelde 'made a marveylous great bombard shotyng stone of marveylous

1. Russell, *Introduction*, pp. xv-xvi.

weyght and when this bombarde shot it made suche a noyse in the goynge as though all the dyvels of hell had been in the way.'[2] In the British Isles, huge 'Mons Meg' appeared sometime during the fifteenth century weighing 14,560 pounds and with a calibre of 20 inches. These weapons were scarcely murderous but were effective in battering down fortresses and city walls. At the siege of Calais (1346) when 'gunnes gret and other gret ordinance' were used,

> gonners to schew ther arte
> in to the town in many a parte
> schote many a fulle gret stone.
> Thankyd be God and Mary myld
> the hurt nothir man, woman ne child.
> To the housis thow they did harm.
> Sent Babara! then was the cry
> when the stone in the stone did fly.[3]

In the open field artillery had only a psychological effect, and even that could not be guaranteed. At the battle of Aljubarrota (1385) the Castilians had sixteen great bombards but the battle was won by the Portuguese, who had no artillery at all.

Guns could be made either of iron or bronze. Iron guns were built by the smith from bars of wrought iron welded into crude tubes which were further strengthened by thick iron hoops shrunk over the tubes. The idea of casting guns must have been there from the very beginning, but iron is very difficult to cast effectively and cast iron is dangerously liable to fractures. Bronze, on the other hand, is technically much easier to cast and all over Europe there were craftsmen well acquainted with the process because of the early and widespread demand for church bells. It is indeed one of the ironies of history that a technique developed in the making of such essentially civilized objects eventually fostered the progress of deadly weapons. Cast bronze ordnance in fact appeared at an early date and met with great favour not only because bronze is less subject to corrosion, but also because

2. Froissart, *Chronicle* ad datam. On Artevelde's huge bombard see Rathgen, *Feuer und Fewiwaffen*, pp. 299–300.

3. From a contemporary poem edited by Wright and Halliwell, *Reliquae*, vol. 2, p. 23. Edward III had at least twenty guns at the siege of Calais: cf. Carman, *Firearms*, p. 23.

the process of casting made the manufacture of muzzle-loaders possible and therefore the avoidance of all the dangers and problems connected with breech-blocks and obturation.[4] In regard to the economic aspect of the alternative, there is no doubt that as raw material iron was much less expensive than bronze. However, until a satisfactory process for casting iron had been discovered, the alternative was between cast bronze and wrought iron. The process of forging guns absorbed more labour and consequently cost more than that of casting. Therefore the final price differential between bronze ordnance and iron ordnance was noticeably reduced.[5]

Copper, the basic raw material for the making of bronze ordnance, came mainly from Hungary, Tyrol, Saxony and Bohemia. Tin, the metal to be mixed with copper, came mainly from England, Spain and Germany. Although the raw materials were produced in only a few areas, the casting of bronze guns was carried on almost everywhere by artisans who had no difficulty in shifting from producing bells to producing guns and *vice versa.* These artisans worked to specific orders or were hired for definite periods of time. Later on, governmental arsenals were also established in which guns were produced by a permanent staff or by experts temporarily hired. It should be added that in those days the division of labour between gunners and gun-founders was not always clear-cut and much casting was actually done by gunners who were serving in the armies.

About the mid-fifteenth century the demand for guns entered a secular phase of rapid growth. The establishment of the great national states with big armies and navies and their incessant wars, together with geographical exploration and overseas expansion, all added to the demand for cannon. The sovereigns

4. Wrought-iron cannon had to be open at either end because it was impossible for the smith to make the barrel properly without a mandrel. The breech-loader was impractical for powerful weapons until the nineteenth century when adequate obturation solved the problem. The difficulties of the detachable breech-block proved insurmountable in big guns. Making the breech-block screw into the breech was not a very satisfactory solution because the heat of each explosion expanded the thread, and for several hours, until it cooled, the breech could not be unscrewed for reloading.

5. Cf. Henrard, *Documents*, p. 243.

themselves became personally interested in the matter of ordnance, and as in the case of Duke Alfonso d'Este, King John II of Portugal, King James IV of Scotland, and the Emperor Maximilian, they developed not only enthusiasm but real technical expertise in the 'art of gunnery'. They patronized gunners and gun-founders and devoted a good deal of their resources to the building and the improvement of arsenals and artillery trains. The trade in raw copper and in bronze ordnance became one of the flourishing and profitable activities of the day, moving mainly around the markets of Nuremberg, the main centre of German metallurgy; Lyons, through which France bought her provisions; Bolzano, on the way from the Tyrol to Northern Italy; and Antwerp where the flow of commodities from West Africa and later the Spice Islands met the flow of metallurgical products from Germany and Flanders. Much of the European *Früh-Kapitalismus* had its origins firmly rooted in this very fertile trade: the Fugger, to quote the most conspicuous example, were great merchants of copper and had a prosperous gun foundry at Fuggerau, near Willbach (Carinthia).

The rapid growth of cannon production was accompanied by technological progress, and attention was now devoted to the guns of lesser calibre. The new trend was advanced by French artillerymen in the course of the fifteenth century. When the army of Charles VIII invaded Italy in 1494 the Italians, then unanimously considered masters in matters of war and artillery, were surprised by the new features of the French ordnance: the huge clumsy proportions of the traditional bombards – as Guicciardini noticed – had been abandoned, the pieces were 'lighter and all cast in bronze ... were drawn by horses with such dexterity that they could keep up with the marching speed of the army ... shot at very short intervals ... and could be used as usefully in the field as in battering walls'.[6] The statement by Guicciardini should be taken with a grain of salt. The new French gun seemed highly 'mobile' in comparison to the old clumsy bombards, but by modern standards they were certainly not conspicuous for mobility. It is generally asserted that at Ravenna in 1512 and at Marignano in 1515 field battles were won by artillery

6. Guicciardini, *Storia*, vol. 1, chapt. 2. Cf. also the more detailed passage by Iovius, *Historiae*, vol. 1, pp. 24–5.

for the first time in history, but one has to admit that other circumstances heavily influenced the result of these battles. European field artillery remained characterized by a low degree of mobility and poor rate of fire until the middle of the seventeenth century. However, by the end of the sixteenth century the distinction between siege artillery and field artillery was recognized and European gunners were devoting their ingenuity to the problem of further improving upon the mobility of guns without affecting their striking power.

Guns were individuals; not only had they individual names like ships, they were actually different one from another. Collado tells us that in the Castle of Milan more than two hundred different types of charging instruments were needed when eleven would have been sufficient had the guns been properly standardized.[7] The first efforts to remedy this state of affairs go back to the sixteenth century and were continued in the seventeenth century, but satisfactory results were not attained before the eighteenth century.

Another problem which became very pressing in the course of the sixteenth century was essentially an economic one, although it had technological implications. We have seen that during the course of the fifteenth century preference was given to cast bronze guns. The trouble with bronze artillery was that it cost too much. Copper and tin were very expensive, and as the need for artillery grew larger, the less bearable became the high cost of bronze. Owing to imperfect technology, the life expectancy of the pieces was very low, and this provided another reason for trying less expensive artillery.

Before the middle of the sixteenth century, the most relevant part of European production, quantitatively as well as qualitatively, came from the Southern Provinces of the Low Countries (Malines, Dinant, Namur, Antwerp, Tournai, Mons, etc.), Germany (Nuremberg, Augsburg, Marienburg, Frankfurt, etc.) and Italy (Venice, Bergamo and Brescia, Genoa, Milan, Naples, etc.). As to quality, Collado says:

There is no doubt among experts that the castings of the Flemish and of

7. Collado, *Platica*, p. 9 r.

the Germans are the best available, and this for various reasons: ... the Germans are phlegmatic people and do things more accurately and more patiently than the Spaniards or the Italians who are choleric people, especially the Spaniards who are the most choleric of all; the Germans enjoy a greater and better supply of copper and tin, with which metals they produce excellent bronze; most important of all the Germans do not use but very dry moulds which they age for years in the sun. ... After the castings of Germany, the castings of Venice are considered very good: in Venice the German style and rules are strictly followed.[8]

While it seems that at least until the beginning of the sixteenth century, Italian production mainly served to satisfy the large local demand for cannon, a good part of the Flemish and German production was exported. In the last decades of the fifteenth and the first decades of the sixteenth century one reads of Flemish artillery sold to England but most of all of German and Flemish artillery sold to Portugal and Spain.

During the last quarter of the fifteenth century Portugal had become an excellent market for cannon merchants. With her involvement in overseas trade and expansion, Portugal's needs for artillery grew vastly beyond her inadequate home resources, while the large profits from overseas commercial ventures translated needs into effective demand. Portuguese kings imported Flemish and German gunners and gun-founders as well as guns; the large imports of copper from Antwerp, even when allowance is made for the metal used in shipbuilding, may indicate that a local industry had been started, but it was of modest proportions and it never grew enough to satisfy the demand. The gold, ivory, and black pepper of West Africa and the spices of the Far East were easily exchangeable in Antwerp for Flemish and German guns. Imports of all kind of weapons were duty free.[9] Thus Portugal remained largely dependent on foreign guns as well as on foreign gunners. The inherent weakness of the situation became obvious in the course of the second half of the sixteenth century, when a severe and prolonged crisis affected the produc-

8. Collado, *Platica*, p. 8 r.

9. Paumgartner, *Welthandelsbräuche*, p. 280: 'all weapons which enter the Kingdom of Portugal are duty free and are not subject to *disma* (decima) nor to *zisa* (assize).' From 1515 copper imports were free from *zisa* but had to pay *disma*.

tive potential of the Southern Low Countries. Throughout the seventeenth century, Portugal suffered from a chronic shortage of artillery that became dramatically acute in moments of crisis when it endangered the safety both of the country and of the empire overseas.

Spain's story is more complicated but substantially not very different. When the country became suddenly and almost unexpectedly engaged in overseas expansion and in European grand politics, its productive capacity in matters of artillery was exiguous. Guns were made in many places by artisans who shifted from the casting of bells to the casting of bronze guns or from the forging of iron utensils to the production of wrought-iron ordnance. This arrangement had been satisfactory in medieval times but it was patently inadequate under the new circumstances. The Crown took steps to face the new situation and built arsenals and gun foundries in Medina del Campo, Malaga and Barcelona. The main problem however was the shortage of skilled labour. 'I do not think,' wrote the Venetian Ambassador Badoer in 1557, 'that there is another country less provided with skilled workers than Spain.' The Spanish administration never managed to break down this formidable handicap. Its action in regard to the manufacture of artillery as in regard to the establishment of a navy remained in general at the level of occasional, short-sighted measures. Mental attitudes of monarchs and officials carried their share of responsibility, but there was more in it than that. The Empire of Charles V had one great asset beside the never-setting sun: it had absorbed within its boundaries most of the European areas with the greatest potential in artillery production. It seemed natural to Spain's rulers that if and when artillery was needed either the orders should be placed to the famous centres of production in Flanders, Italy and Germany; or Flemish, German and Italian founders should be invited to Spain. The second alternative was frequently adopted and if consistently pursued would have undoubtedly helped to solve Spain's problem. But after the emergency needs were satisfied, the foreign gunners were sent back to their homes for economic reasons and the few Spanish workers were left without jobs or money. In the short run the action of the Crown could hardly be condemned. Its failure was in the long run.

During the second half of the sixteenth century, wars, religious intolerance and misgovernment disrupted the economy of the Southern Low Countries. Skilled labour emigrated *en masse.* Italy did not keep up with contemporary technological developments and proved unable to fill the gap left by the Flemish crisis. With the collapse of two of its main suppliers, Spain underwent an acute armament crisis. The crisis is already apparent in the 1570s and it grew alarmingly worse in the following decades[10].

A few initiatives did result in substantial and durable achievements, such as the establishment of the royal gun foundries in Seville in 1611 and the establishment of cast-iron manufactures in Biscay in the 1620s, but on the whole Spain's armament industry lagged behind the needs of Spain's grand imperial politics. The country had often to depend on the enemy's manufactures.[11] For this, the Crown and its bureaucracy were not the only ones to be blamed. In contrast to English, Dutch and Swedish developments, private enterprise in Spain was conspicuous for its inaction, and not only in the armament industry. 'Spain supplies itself from other countries with almost all things which are manufactured for common use and which consist in the industry and toil of

10. From the 1570s onward one frequently hears of frantic efforts on the part of Spaniards to secure imports of English cannon. In 1578 'a person' offered to get 150 guns out of England and deliver them to Spain. The whole transaction had to be made in great secrecy '*para que no se sepa en Inglaterra donde hai tanto recato en la saca de artilleria*'. In 1583 twenty-three guns were imported from England for a total weight of 13·5 tons of iron besides 1·5 tons of ammunition (Carrasco, *Artilleria de hierro*, p. 66). In 1588 two Spaniards attempted to induce an Englishman, Richard Thomson, 'to undertake the delyverie of a great quantitie of yron ordnance, either in Hamborghe, Rotterdam in Holland or Callis, profferinge to deposit in any of these places twentie thousand crownes to receave the valiewe in cast ordnance' (*Calendar State Papers*, Domestic, Elizabeth, 244 *n*. 16, April 3, 1593). So urgent was the demand that the Spaniards offered him a pension of forty ducats for life. In 1591 the Spaniards tried again to import iron cannon from England (Cunningham, *English Industry*, vol. 1, p. 57). In 1619 'great quantity' of English ordnance was imported into Spain, through legal channels (Cal. St. Pap., Domestic, 13 Feb. 1619).

11. As indicated above in footnote 10, the Spaniards imported iron cannon from England during the last decades of the sixteenth century and the first decades of the seventeenth century. They also imported ordnance from the Netherlands throughout the seventeenth century (Janiçon, *Provinces-Unies*, vol. 1, p. 498).

man', wrote a well-informed contemporary observer; and the Venetian Ambassador Vendramin, commenting on the effects of this situation on Spain's balance of payments, added: 'This gold that comes from the Indies does on Spain as rain does on a roof – it poures on her and it flows away.'

Across the Channel, in the early years of the sixteenth century England looked almost insignificant. There were less than four million people living in England and Wales, while France certainly numbered more than ten million and Spain about seven million in addition to another four million in its newly conquered Italian territories. The small size of the English population was not offset by greater *per capita* incomes or productivity. On the contrary, to any continental observer England looked most certainly an underdeveloped country. The English cannon industry and English armament were no exception. The fact that the bow was not finally and officially discarded as an English weapon until 1595 belongs more to the history of English folklore than to a serious history of English technology. But even when due allowance is made for folklore and tradition, it remains substantially true that during the fourteenth and fifteenth centuries the English armament industry did not keep pace with that of the Continent. One reason was that the English had a good supply of easily found iron ore in the proximity of well-forested lands. This made them stick mostly to iron in their metallurgical endeavours, and put them at a disadvantage as long as the technique of working iron remained less developed than that of working cupreous alloys.

One of the main areas of the English iron industry was in the Weald of Sussex, in Ashdown Forest. Here in the last decades of the fifteenth century, under the impulse given by Henry VII, the production of artillery expanded. Much effort was devoted to improving current techniques, and technological progress was undoubtedly achieved. From available records it seems that many of the founders who worked in Ashdown Forest were gunners of French origins which proves that even in the field of iron work the need was felt to import technicians and 'know-how' from the more developed continent. Between 1490 and 1510, side by side with the casting of gun-shot, the casting of iron guns was tried and by 1509–13 we know that guns made of cast iron were manufactured. We have no way of telling how good such artillery was, but it is unlikely to have been very satisfactory.

When Henry VIII (1509–47) came to the throne, he was quick to realize his weakness in the matter of artillery. The majority of the English naval and land guns were of the old outdated type, constructed by welding together longitudinally a number of wrought-iron bars. There was only one bronze gun foundry in England, the one at the Tower, and this could not possibly cope with the King's vast needs. The cast-iron guns made in the Weald apparently did not inspire him with much confidence. Full of admiration for continental bronze guns, Henry, like the Spanish monarchs, turned to the famous gun foundries of the Southern Low Countries, and placed large orders to the great master Hans Poppenruyter of Malines. In less than two decades he received from Poppenruyter at least 140 bronze guns of all calibres including the twelve great bombards named 'the twelve Apostles'.

At the beginning of his reign, King Henry was able to keep pace with his extravagant expenditure thanks to the treasure that his father had carefully collected, but before long this was exhausted and Henry did not have at his disposal the American resources that the Spanish monarchs had. As early as 1523 Master Poppenruyter was – to use his own words – 'in great dispair and danger' for the money due to him by the King, and after several vain attempts to obtain payment he declared himself ready 'to make finance with artillery', which possibly meant that he was ready to reach a settlement.

When war with France became imminent in 1543 and armaments had to be increased, Henry was practically bankrupt and the need to resort to native industry and local raw materials became dramatically obvious. It was Henry's good luck that in Ashdown Forest, iron-founding had never ceased completely although much neglected. Only two years before, in 1541, Henry had appointed William Levett subtenant of the royal iron works at Newbridge. William Levett was the rector of the parish of Buxted and although he had distinguished himself as a deputy for the receiver of the King's revenues in Sussex, the King's choice might look rather odd. But it turned out to be an excellent one. In 1543, when the King needed cannon, the enterprising parson secured the services of a large number of French founders who were working in the vicinity, obtained from London Peter Baude, the best founder of bronze guns in the King's service, and completed the group with a skilful iron-master Ralph Hogge who knew how to operate a furnace and was able to furnish the liquid iron for the castings. Having thus collected the best experts available, the energetic parson set them to work and somewhere in his parish a number of good cast-iron guns were successfully produced.[12] The event marked the beginning of a prosperous era for the iron industry in Sussex and opened a new chapter in the history of artillery.

The new guns gave such good results that in 1545 Levett was ordered to cast no fewer than 120 iron guns which he managed to produce in nearly two years.[13] The further request for large siege guns led to the building of a double furnace in the Forest of Worth, west of Ashdown Forest, which was the first furnace of that kind in England known to us.[14] By 1573 there were eight furnaces in Sussex and one in Kent casting guns and shot for a total annual output of 500–600 tons of iron. Around 1600 annual production had increased to about 800–1000 tons.[15]

The reasons why the English should have almost suddenly succeeded in the production of relatively safe cast-iron guns is not completely clear. Whatever the reasons for the technological success, we know that it turned into an economic one. As has been rightly pointed out, 'the manufacture of guns was the most profitable proposition in the sixteenth century iron trade'. The production of cast-iron guns grew very rapidly while the fame of English products and skill quickly spread throughout Europe. Before the century was over, an expert like Gentilini, a Venetian gun-founder of good repute, although strongly biased in favour of Continental artillery, wrote that 'English people, to say the truth, are judicious people and of great intelligence . . . and are very ingenious in their inventions.'[16] By then, English technicians as well as English guns were eagerly sought after throughout the

12. Schubert, *Iron Industry*, pp. 171–2.
13. op. cit., p. 247.
14. ibid.
15. op. cit., p. 250.
16. Gentilini, *Bombardiere*, p. 55.

Continent. To use Toynbeean phraseology, it was a typical case of challenge met with bold and successful response.

Actually cast-iron guns were in many respects still inferior to bronze artillery. This was certainly by far the most prevalent opinion among military experts, especially on the Continent, where cast bronze artillery had reached a high state of perfection. 'Brass ordnance does not become rusted,' wrote Gentilini, 'and therefore can be fired without danger for the gunner, while this is not the case with iron ordnance because shots become rusted when they have been inside the gun for a while.'[17] But there was more to it than that. Despite the enormous progress accomplished by English founders, cast-iron guns were still more brittle than bronze ones. In the Ordnance Minute Books in the London Record Office one reads over and over again notes of cast-iron guns that missed proof, and the records of accident in action abundantly testify that cast-iron guns were less reliable than cast bronze artillery. Furthermore, because of the lower resistance of the metal, iron guns had to be made noticeably thicker than bronze ordnance. As a result, cast-iron guns were much heavier than corresponding bronze guns.

However, cast-iron guns had one great advantage over cast bronze ordnance: that of costing much less. The price of bronze artillery seems to have averaged three or four times that of cast iron ordnance.[18] Essentially, therefore, the English had come out

17. ibid.

18. Iron guns had to be made much heavier than the bronze ones for safety reasons, thus the differential cost between similar pieces was somewhat reduced. However, in England around 1632, John Browne presented a plan for the substitution of cast-iron ordnance for bronze ordnance on all the smaller ships of the Navy. Browne stated that ninety tons of bronze ordnance would cost £14,332 5s. but if made of iron would cost only £3,600. Around 1636 it was estimated, always in England, that the cost of casting culverins and demi-culverins weighing 33 tons 12 cwts would have been £5,355 2s. in bronze but only £1,176 in iron (*Calendar of State Papers*, Domestic, 1631–3, vol. 230, *n.* 36 and 1636, vol. 340, n. 48). In 1578 in Spain '*el Capitan General del Arma*' Don Francis de Alava emphasized 'the great utility of introducing into the Kingdom' the production of cast-iron cannon and pointed out that while bronze cannon cost 16 ducats per hundred kilos weight, cast-iron cannon cost only 5½ (Carrasco, *Artilleria de hierro*, p. 66). In 1644 Louis de Geer told the Swedish Privy Council that cast-iron cannon cost one third the price of bronze cannon (Heckscher,

with a product whose deficiencies in quality were more than compensated by advantages in price. In their contemporary struggle to outsell foreign competitors in the textile sector, the English showed exactly the same attitude and followed the same line of action. Their national genius for practicality becomes lucidly obvious in matters of decoration. 'Through the whole series of English guns,' wrote Mr Ffoulkes, Master of the Armouries at the Tower of London,

> from the reign of Henry VIII to the middle of the nineteenth century one of the noticeable features is the simplicity of form and subordination of decoration to practical needs. ... The same realization of technical and practical needs is to be found in English armour throughout the ages. We were never caught up in the maelstrom of magnificent incoherence which marks the designs for armour by Cellini, Campi, Giulio Romano and the Louvre School.[19]

What mostly mattered to the English was that their products would perform the task and cost as little as possible. No more striking contrast can be found to this attitude than the extravagance of some contemporary Italians who for the sake of beauty were engraving and decorating not only the guns but even the gunshots, knowing perfectly well that this was detrimental to the efficiency of their artillery.

By 1567 Queen Elizabeth granted Ralph Hogge the monopoly of exporting 'cast iron ordnance with gunstone', but already by 1573 Hogge was complaining that his privilege was constantly being infringed and that other gun-founders were exporting to Sweden, Denmark, France, Spain, Holland and even Flanders.[20] The structure of the demand for English cannon was then undergoing remarkable changes. Because of the increase of English privateering and seaborne trade, private internal demand grew very rapidly, and in time of peace it was much more substantial than public demand: in 1621 John Browne declared that the King's service would have only taken up ten days a year of

---

Ekonomiska Historia, part 1, vol. 2, p. 454).

19. Ffoulkes, *Gun-Founders*, p. 28.

20. Tawney–Power, *Documents*, vol. 1, pp. 262–3. The report by Mr Hogge gives a colourful account of the way in which guns were smuggled out of the country. For exports to Denmark cf. Christensen, *Historia*, pp. 24–5.

his furnace's activity.[21] Foreign demand also became increasingly relevant, and around 1573, according to Master Hogge, 'there is above 400 tonnes cast yearly [in England] and all this will not be sold nor bought to remain within the Realme.'[22] This aroused suspicion and anxiety among politicians. English guns were, in the words of Sir Walter Raleigh, 'a jewel of great value'; why, argued the politicians, should the English place such 'a jewel' in the hands of potential enemies? This point of view was largely shared in political circles, and as a result Queen Elizabeth issued an order in 1574 restricting the number of guns to be cast in England to those 'for the only use of the Realm'. From that time onward the export of guns was a point of contention between gun-founders anxious to sell as large a number of guns as possible whether at home or abroad and politicians no less eager to prohibit this export altogether.

Licences could be obtained to export ordnance, especially (although not exclusively) to friendly Protestant powers. In 1619 Thomas Browne, who owned probably the largest gun foundry in England, admitted that half of his production was exported under licence to Holland,[23] 'the Dutch havinge bargained with him to

21. *Calendar State Papers*, Domestic, Addenda 1580–1625, vol. 42, no. 66, dec. (?) 1621.

22. Tawney–Power, *Documents*, vol. 1, pp. 262–3. Some calculations preserved in *Calendar of State Papers* 14, vol. 26, no. 52, 17 February 1607 seem to suggest that between 1596 and 1603 about 2,270 tons of cast-iron ordnance were exported from England under licence (about 58 tons by English merchants and 2,212 tons by aliens), at an average of about 325 tons per year. The figure sounds reasonable, although the text of the document is obscure and its interpretation open to dispute.

In 1601 a certain Edward Peake made an appeal to the House of Commons to forbid any export of ordnance. He maintained that the Queen might make £3,000 a year out of a custom of £4 a ton on exported ordnance. He therefore implied a volume of exports as high as 750 tons per year (Taylor, *Camden's England*, p. 357). However, in the *Calendar of State Papers* 14/8/132 (July 1604) one reads that 'the impost of ordnance ... may be valued at about £1,200 a year.' At the rate of £4 per ton, this would bring us back again to about 300 tons per year.

These were legal exports. To them one should add the illegal ones, whose amount is of course totally unknown.

23. *Calendar of State Papers*, Domestic, 1619–23, vol. 105, no. 92, February 11, 1619.

take of all that the English doe not buy'.[24] When licence could not be obtained evasion was tried. However, despite licences and surreptitious sales, there is no doubt that exports were hampered by the very existence of the general prohibition.

The views of the politicians were not absurd, but it was totally unrealistic to believe that the Continent, with an abundant supply of entrepreneurial ingenuity, and manufacturing skills, would easily submit to English superiority, especially in such a vital matter as the armament industry.

In 1574 there was a state of chronic war in the Low Countries between the Spanish-Catholic forces on one side and the Protestant dissidents on the other. Don Luis de Requesens, the new Spanish governor, was in principle favourable to the idea of peace negotiations but faced with a rapidly deteriorating military situation he had first to try to reorganize his forces and strengthen their armament. He revitalized the royal gun foundries at Malines by sending there about 35,000 lb. of copper from Hungary and about 2,000 lb. of English tin to be made into 38 guns.[25] But copper and tin were expensive, Requesens's financial situation was at the point of bankruptcy and he needed more than 38 new guns. It is, therefore, not at all surprising if in the midst of his financial troubles the governor thought of English iron ordnance as a workable solution and placed a large order for cannon directly in England. Those, however, were the days when Queen Elizabeth had prohibited the export of English guns especially to Catholic powers, and no licence was issued in favour of Don Luis's request. Faced with the English refusal, the governor turned to continental foundries. Liège, then the capital of an independent and uncommitted principality, could boast of a great tradition in iron works since the Middle Ages. Guns had never been made there, but iron gun-shot was currently cast and light arms were also produced. Requesens contacted a local

24. Schubert, *Iron Industry*, p. 249. Not many years before, in 1612, Thomas Browne was in trouble for having 'transported much iron ordnance from that furnace (at Brenchley) against the law restricting the export of such'. The justices of the peace were ordered to seize all the iron ordnance stored at furnace (Schubert, *Extension*, p. 246).

25. Henrard, *Documents*, p. 258.

prominent manufacturer, Wathier Godefrin, and placed an order for 46,000 cannon balls and 300 guns to him: a grand total of 620,000 lb. of cast iron. Time and place of delivery: within six months in Antwerp. Despite the relatively short time allowed, guns and ammunitions were cast and delivered on time, but, alas, at the testing, the guns failed. The disappointed Spaniards got hold of M. Godefrin and clapped him for a while in the jails of Antwerp.[26]

The list of continental failures in the race with English technology did not end with the misfortunes of M. Godefrin. A Spanish project to introduce the new technology in Biscay in 1574 failed because the available founders from England of Liège were not willing to move to Spain for fear of the Inquisition '*en que todos estaban harto danados*'. In 1603 Flemish masters were again taken to Spain in order to set going the production of cast-iron guns, but again the project had to be abandoned with a heavy financial loss for the administration. However, in the course of time, it was English politicians and not English gun-founders who were proved to be wrong.

The effective catalyst for continental developments was Holland. The Dutch needed cannon in ever-increasing quantity for their incessant wars against the Spaniards, for their huge navy, and for their overseas expansion. Their rapidly increasing wealth transformed need into effective demand. On the other hand, the conflict with Spain cut off Holland from the centres of cannon production in the Southern Low Countries. The Dutch had to find new solutions for their new problems. English cast-iron guns seemed at one point to offer a practical way out. Between 1560 and 1600 the Dutch imported large quantities of them[27] and looking back over the last decades of the sixteenth century a Venetian Ambassador wrote in 1603 that the Dutch 'had been able to supply themselves from England with all necessities and especially with artillery'.[28] But in 1574, under the pressure of

26. For all that precedes cf. Lejeune, *Capitalisme*, p. 185 and Evrard and Descy, *Vennes*, pp. 41–2.

27. Elias, *Zeewezen*, vol. 1, p. 56; Baasch, *Wirtschaftsgeschichte*, p. 270*n*.

28. Barbour, *Amsterdam*, p. 36, *n*. 93.

English politicians, Elizabeth had established a control on the ordnance trade. Licences were more freely issued for exports to Holland than to the other countries. Moreover, surreptitious exports to Holland were so frequent and easy that they caused much loud complaint in England. But the arrangement was not very satisfactory for the Dutch. Furthermore, with the beginning of the seventeenth century English industry ran into a serious bottleneck. As we shall see later, this must have affected English exports negatively and the first to suffer the consequences were obviously the Dutch. Guns were always in short supply in the United Provinces, and when the navy needed to equip the larger ships of the fleet for an action, guns had to be taken from the battlements of towns.

Under the pressure of the circumstances the Dutch set on foot a local cannon industry. Gun foundries, private as well as public, sprang up like mushrooms in Maestricht, Utrecht, Amsterdam, Rotterdam, The Hague and by the early seventeenth century the Dutch looked to Fynes Moryson 'no less witty than industrious so particularly they have great skill in casting great ordinance'.[29] In an early phase they kept to bronze artillery but they did not spare efforts to emulate the English in casting the cheaper iron guns. In 1601 and 1619 patents were granted to residents in Holland for the casting of iron guns in the English way.[30] In 1604 a double furnace was operating at Asslar west of Wetzlar in Germany, where iron ordnance was cast mainly for the Dutch.[31] In the 1620s iron guns were cast by the Dutch at Marsberg in Westphalia.[32] By then the new technique was spreading throughout Europe, also in areas that were not under direct Dutch influence.

In the course of time the Dutch arranged their armament production on a double basis. They retained the production of bronze guns in the Netherlands, where, through their commercial network they could amass copper from Sweden and Japan and tin from England and Germany. They organized the casting of

29. Moryson, *An Itinerary*, (Glasgow's ed.) vol. 4, p. 474.
30. Doorman, *Patents*, pp. 101 and 118.
31. Schubert, *Cannon*, p. 138 P.
32. Van Dillen, *Amsterdam*, doc. 537 and doc. 539; Doorman, *Patents*, p. 118.

iron guns abroad where suitable iron ore and charcoal were available.

Sweden is naturally endowed with abundant and excellent copper, tin and iron ores, with vast charcoal-producing forests and enough streams for water power and transportation. Since the days of Gustav Vasa (1523–60) manufactures for the production of fire-arms were being set on foot, especially in the mining districts of Central Sweden. A very active role in these developments was played by the Crown, which owned many of the factories and tried to improve the standard of production by employing foreign technicians.

Three phases seem to stand out: a first between 1530 and 1560 in which a number of factories were opened for the production of wrought-iron guns; a second, between 1560 and 1580, in which foundries were established, for the production of cast bronze ordnance; and a third after 1580, which saw the appearance and growth of the furnaces for the production of cast-iron guns.[33] In catching up with the rest of Europe, Sweden was concentrating into a few decades the evolutionary process that the Continent had taken centuries to accomplish.

33. Jakobsson, *Beväpning*, pp. 44–6, gives a list of gun factories that operated in Sweden in the sixteenth century. For each factory he gives either the date when gun production was started or the date when the first mention of arms production is encountered in the extant documents. I have arranged the data in the following table. Jakobsson does not make reference in his list to the gun factory at Åkers; I have therefore integrated his information with the data that I derived from Hahr, *Åkers*, p. 6. There are some other gaps in Jakobsson's list that I could not fill, but I do not think that the picture could be substantially changed.

Year in which the factory began to operate or first mention is made of its cannon production	*Number of factories producing*		
	Wrought-iron cannon	*Cast bronze cannon*	*Cast iron cannon*
1530 to 1540	2		
1541 to 1550	5		
1551 to 1560	2	2	
1561 to 1570	2	2	
1571 to 1580			1
1581 to 1590		1	3
1591 to 1600		1	4
1601 to 1610			3

Cast-iron guns in Sweden are first heard of in 1560, but it seems that this was only a matter of more or less experimental activity. The effort was evidently continued and after 1568 there are no further references to wrought-iron guns, while there is clear evidence of castings of iron artillery. How satisfactory such cast-iron guns were, it is difficult to say, but it seems that they were not very good. Production seems to have been very limited also from a quantitative point of view. The fact of the matter was that the growth of Swedish manufacture was severely hampered on the one hand by shortages of capital, entrepreneurship and skilled labour and on the other by the lack of financial and commercial organization to help in the marketing of the product abroad.

Knowing how desperately the Dutch were looking around for ordnance, it cannot be surprising if they soon became interested in the Swedish developments. The growth of their trade and activities in the North, and the fact that some technicians from the Low Countries were working in Swedish factories, undoubtedly helped to the establishment of the contacts. The arrival of the Dutch brought to Sweden the human element and the capital necessary to activate Swedish wealth fully. At the same time the Dutch commercial organization transmitted to the Scandinavian manufactures the stimuli of the large demand for cannon connected with the Dutch and English overseas expansion, the Dutch–Spanish wars and the Thirty Years' War. Within the first three decades of the seventeenth century, Swedish industry underwent a revolutionary change and Sweden took its place in the front rank of the European armament industry.

The key men in this phase of our history were Guillaume de Beche, Elias and Jacob Trip and their sons and especially Louis de Geer, the fabulous founder of a huge industrial empire. The Dutch were essentially interested in getting cast-iron guns from Sweden and the nature and weight of their demand strongly influenced the patterns of Swedish developments. The production of cast-iron guns expanded very rapidly and so did exports. We first hear of exports of some size in 1615 when a letter of Gustavus Adolphus discloses that the States General of Holland had requested 400 Swedish iron guns.[34] In 1620 an unknown quantity of

34. Heijkenskjöld, *Styckegjutning*, pp. 72–3.

guns was purchased by Elias Trip and shipped to Holland.[35] In the same year it was noticed in Amsterdam that Swedish cannon were finding an easy outlet there.[36] Six years later exports of cast-iron guns from Sweden amounted to nearly 22 metric tons.[37] From then onward, they increased at a great speed. Over the period 1637–40 they reached the level of *c.* 780 metric tons per year, between 1641 and 1644 *c.* 940 tons and in the three-year period 1645–7 *c.* 1,100 tons per year. Between 1655 and 1662 the yearly average was again around the 1,100 tons, with a peak in the years 1661 and 1662 when exports reached respectively 1,459 and 2,556 tons (see Table 1) mostly shipped to Holland.[38] As to the numbers of guns involved, it may be calculated that the average weight of the pieces exported varied in different years, ranging generally between *c.* 610 kg. (in 1660) and 1,810 kg. (in 1662). The number of pieces exported in selected years was as follows:[39]

1655	367	1660	1,150
1656	1,048	1661	2,440
1657	698	1662	1,412
1658	1,191		

35. ibid., p. 75, *n.* 1.
36. Dahlgren, *De Geer*, vol. 1, p. 115.
37. Cf. Table 1. It does not seem that 1626 was a year of exceptionally low exports. By that time the Swedish cannon industry was not yet fully developed. In a letter written in 1626 from Lübeck, Conrad von Falkenberg noticed: 'Swedish iron cannon and ammunition are much in demand and I could do good business if only I had enough to sell'. Two years later von Falkenberg wrote from Amsterdam that 'the guns that come from Sweden are considered very good. The people here use them to arm their vessels and do not go very much after other artillery. There is no evidence that many iron-guns are now imported from England.' (cf. Heijkenskjöld, *Styckegjutning*, p. 73).
38. Not all cannon exported to Amsterdam was necessarily retained in Holland, much being re-exported to England, Portugal and other places.
Direct exports of cast-iron cannon from Sweden to England are first mentioned in 1632.

In 1694 iron cannon was exported from Sweden to the following places:

England	10 metric tons	Denmark	36 metric tons
Holland	8	Hamburg	65
Portugal	114	Lübeck	48
Wismar	1	Rostock	28

To give a meaning to these figures, one can say that the guns exported in one year from Sweden were enough to equip a small fleet, or at least half a dozen powerful squadrons.

Table 1

Exports of Cast-Iron Cannon from Sweden
(*Metric tons*)

Year	*Total*	*Of which to Holland*	*Year*	*Total*	*Of which to Holland*
1626	22		1656	1,234	1,000
1637	576		1657	778	683
1638	467		1658	1,242	1,131
1639	1,047		1659	243	
1640	1,044	1,044	1660	931	873
1641	1,202	1,202	1661	1,459	1,277
1642	1,156		1662	2,556	2,394
1643	654				
1644	761		1664	1,274	
1645	1,498				
1646	1,084		1668	1,346	
1647	728				
1650	1,210		1685	259	
1655	364	364			
			1694	432	8

Source: Heijkenskjöld, *Styckegjutning*, pp. 73–4. The figures are originally given in *skeppund*. The value of this unit of measure varied from place to place, from about 195·4 kg. in some districts to 136 kg. in most 'staple districts' (cf. Jansson, *Måttordbok*, pp. 72–3). According to the indications given by Hildebrand, *Historia*, p. 457, I have adopted the equivalence of 136 kg. (1 metric ton = 1,000 kilos = 0·9842 English tons = 7,353 skeppund).

The figures given by Heijkenskjöld correspond to the figures derived from different sources and published by Boëthius-Heckscher, *Handelstatistik*, pp. 102–3 for some years (1637–41, 1645 and 1650). Only for the year 1641 does one notice a small difference between the two sources (8,840 skeppund according to Heijkenskjöld and 9,293 according to Boëthius-Heckscher). For a number of years no data are given in Table 1. This is imputable to the state of the available Swedish records and it does not imply that in the

---

Danzig	52 metric tons	Nyenskans	12 metric tons
Riga	14	Narva	41
Reval	2		

All data are derived from Heijkenskjöld, *Styckegjutning*, pp. 75–6. As to the metrological problems involved cf. above in the text note to Table 1.

39. ibid.

corresponding years there were no exports of cannon from Sweden. The documents relating to the passages at the Sund (Bang, *Tabeller*, ad annum) prove that iron cannon was brought 'westward' also in the years 1622, 1623, 1624, 1625, 1627, 1628, 1629, 1630, 1631, 1633, 1635, 1636, 1648, 1649, 1651, 1652, 1653, 1654, etc. The absence of any figure in the column *Holland* means that indications are not given about the destination of Swedish exports for the corresponding years.

Germany was another area where Dutch demand for cannon gave impetus to the manufacture of iron artillery. Mention has been made of the iron works at Asslar, which were operating as early as 1604. In 1612 the Dutch controlled two of the six blast furnaces at Marsberg in Eastern Westphalia, where in the 1620s they set on foot iron cannon production.[40] The Dutch were also the main customers of Jean Mariotte, a native of Liège, who between 1630 and 1650 established successful manufactures of iron ordnance at Weinhär, in the domains of the Abbey of Amstein, and at Stromberg, north of Koblenz.[41]

The Thirty Years' War, while stimulating local demand for artillery, caused widespread destruction of plants and loss of skilled labour. Developments were therefore erratic. But after the middle of the seventeenth century, the production in Western Germany grew very rapidly and by the 1660s Swedish industry was feeling its competition. It seems that the main centre of trading in German iron ordnance was Cologne, where the Dutch were active buyers still at the end of the century.

The Dutch were also instrumental in developing gun foundries in Russia. In the 1630s a Dutch group established iron works near Tula, about 120 miles south of Moscow[42] and allegedly was the first to bring to Russia western and modern methods of casting iron.[43] Skilled labour was not available in Russia, and technicians had to be imported from abroad. This was not easy. On the other hand, with serfdom still prevailing in the country, unskilled labour was plentiful and could be secured at very low cost. It was largely employed in cutting trees and in providing fuel. The government looked very favourably upon the Dutch initiatives

40. Schubert, *Superiority*, p. 86.

41. Yernaux, *Métallurgie*, pp. 79 and 172–5.

42. For the story of the enterprise cf. Strumilin, *Istoriia*, pp. 102*ff*, and Amburger, *Marselis*, pp. 92–130, 150–54 and 168–74.

43. Portal, *L'Oural*, p. 188.

and supported them with annual subsidies as well as with grants of rights on serf-labour.[44] But the result of all these efforts was not very satisfactory. The products of the Russian foundries could be offered at a very low price in Amsterdam, but their quality was extremely poor, and still at the end of the seventeenth century Russian production was not considered very important.[45] Anyhow a nucleus had been established and its significance was to become clear in the following century.

The early imitations of English cast-iron guns were far from satisfactory: 'Att first,' said an English Report dated 1623 about Swedish products, 'theis peeces in the proovinge brake most of them'[46] and in 1627, out of eleven cast-iron guns that the Dutch obtained from France, six exploded during 'reasonable testing' and one lost its rear.[47] But soon substantial improvements were accomplished and by 1623, according to the English Report, Swedish guns were 'so amended that they are scarce to be knowne from English peeces and now keepe good proofe'.[48] That they had been 'amended' and that they kept 'good proofe' was certainly true, but to say that 'they are scarce to be knowne from English peeces' was possibly an exaggeration in which the

44. Strumilin, *Istoriia*, pp. 106*ff*, and Mavor, *Russia*, vol. 1, pp. 434–5.

45. The Marquis de Seignelay wrote in 1671 to his father Colbert that the Dutch were importing 'a few cast-iron guns from Moscovia which however are not very good and are sold at very low price' (Clément, *Lettres*, vol. 3, part 2, p. 311). In his *Addition au memoire concernant la fonderie des canons* dated July 1671 (Bibl. Nat. Paris, Dept. MSS., Colbert 4219, *f.* 11) the Marquis added: '*à les gard des canons de fer il ne s'en fond point en aucun endroit de Holande et il vient tout fondu de Suede et de Moscovie quoy qu'il en vien assez peu de Moscovie. . . . . Le fer de Suede est le meilleur et le plus estimé. Il en vien peu de Moscovie qui mesme n'est pas bon*.' In 1674 Kilburger (*Handel*, p. 324) wrote that the cannon produced at the Russian factories (only up to 24-pounders) 'were exported via Archangel to Holland where they generally blew up at their testing'. According to Amburger (*Marselis*, p. 105) the poor quality of the Russian products was mainly due to shortage of skilled labour. On the report by Kilburger and its information on Russian cannon cf. also Kurts, *Socinenie Kilburgera*, pp. 451–69. On the report by Kilburger in general cf. Nyström, *Mercatura*, pp. 239–96. According to Strumilin (loc. cit.) 600 cannon were exported to Holland in 1646 and 360 cannon in 1647.

46. Public Record Office, London, State Papers 14/155 no. 11 (*f.*840), 3 December 1623: 'Relation of Iron Ordenance made beyond the Seas'.

47. Van Dillen, *Amsterdam*, doc. 1108.

48. See above footnote 46.

author of the Report indulged in order to get help from the Crown for the local industry. The son of Colbert, the Marquis de Seignelay, reported to his father in 1671 that 'there is a great difference between the Swedish and the English guns, those cast in England being far superior'.[49] This also was possibly an exaggeration in which the Marquis indulged under the influence of English current opinion, but it is not improbable that he was nearer to the truth than the author of the English Report of 1623. The Dutch were not so concerned with quality as they were with costs and in their trades they were always ready to sacrifice the first to the latter. They did so also in the cloth trade and in the wine trade. If they did not produce iron guns as good as those produced in England, they certainly succeeded in underselling English iron ordnance and in making Amsterdam the main munition market of Europe. Their task was made easier by England's increasing difficulties with her local supply of timber. Since the accession of Elizabeth complaints about deforestation were frequently heard and in the course of time became more alarmingly frequent in one county after the other. In Table II, I have compared a general price index with an index for the price of charcoal in England during the period 1560 to 1670. The results of a comparison between the two series should be accepted with reserve. General price indexes look attractive when neatly printed but their real meaning is always obscure. The series of charcoal refers to one limited area only and other instances should be considered before drawing general conclusions. Tentatively, however, one may say that according to available figures the fuel crisis seems to have exploded in all its gravity during the 1630s. English iron guns were still exported in the second and third decades of the seventeenth century but, notwithstanding the vociferous allegations of patriots, it does not seem that the trade was very prosperous. In the 1630s when the fuel crisis became acute England began to import iron cannon. The first mention one encounters of Swedish iron guns shipped directly from Sweden to England is in 1632[50] and from 1638 we

49. Clément, *Lettres*, vol. 3, part 2, p. 332.

50. See above footnote 38, p. 54. In the Sund's statistics one also notices iron cannon being brought 'westward' on English vessels in 1647 and 1648 (see Bang, *Tabeller*, ad annum).

know of many shipments of Swedish guns from Amsterdam to England.[51] In the early 1670s the Marquis de Seignelay wrote that the English 'not having enough timber for casting all the ordnance they need, get cannon from Sweden although they esteem that the Swedish iron is not as good as that of England'.[52]

Table II

Indexes of Prices in England
1560–1670 (1630=100)

	price-indexes	
year	*general*	*charcoal*
1560	46	60
1610	90	95
1620	87	100
1630	100	100
1640	106	135
1650	133	225
1660	121	220
1670	102	250

Sources: Phelps Brown and Hopkins, *Prices*, pp. 194–5 and Beveridge, *Prices*, pp. 144–5. In order to avoid the effects of the short run fluctuations of food prices, the figures for the general price index have been calculated on the basis of a five years' average about the corresponding year.

The use of cast-iron guns became progressively more common in the course of the seventeenth century especially on the seas, and by the end of the century cast-iron guns were the predominant artillery aboard European vessels.

It was mainly in connexion with the needs of the navy that Colbert took action to remedy the weakness of France in the armament industry and to develop the manufacture of cast-iron guns. After the exploits of French artillerymen during the period 1450 to 1550, the French cannon industry entered a period of disruption and decline. The civil wars and the political confusion that followed were mainly responsible for this state of affairs. A chaotic administration, weakened by antagonistic factions, could not lend any organized support to the armament industry, while this could not live without some kind of government assistance.

51. Barbour, *Amsterdam*, p. 38, *n*. 98.
52. Clément, *Lettres*, vol. 3, part 2, p. 322.

Many skilled workers left the country for religious reasons or in search of better pay and more security. Amid troubles and wars, paradoxically enough, the cannon industry collapsed. A few manufactures survived here and there but on the whole they were of no importance. For armaments France became largely dependent on foreign supply.

The reconstruction ruthlessly and skilfully accomplished by Richelieu was not only limited to the political and administrative sectors. The energetic Cardinal-Duke rebuilt the French army and remade the French navy practically from scratch. But he did not rebuild the French armament industry. A few timid attempts to develop cannon manufactures at Brouage and Le Havre were of no great consequence.[53] Richelieu remained essentially dependent on foreign supply. His market was Amsterdam and there he had permanent agents who bought large quantities of bronze and iron ordnance, muskets, anchors and gunpowder.[54]

After the tight-budget years of Mazarin, the rearmament policy was vigorously revived by Colbert. French productive capacity however was still almost non-existent in the early 1660s. In regard to the iron industry there had been some modest developments in Anjou, Normandy and Brittany during the 1630s, but the Thirty Years' War had ruined the manufactures in Lorraine and Champagne while the civil war at the time of the minority of Louis XIV had provoked new waves of emigration of skilled workers and the closing of many foundries all over the country.[55] The small isolated furnaces with very low capacity were the typical production units of the French iron industry. The dearth of capital was sorely felt. Saving was not lacking but the nobility and the Church were not interested in industry and the 'Third Estate' liked to invest in '*les charges et les offices*'. In striking contrast with Dutch and English developments, private enterprise was conspicuous for its inertia. The propensity of the government to buy cannon abroad did not help to change this state of affairs. On the other hand, since local production was inadequate, the government was obliged to turn to foreign producers. It was a vicious circle. From 1661 to about 1666

53. Basset, *Historique*, p. 989.
54. Basset, *Historique*, pp. 988–9 and Barbour, *Amsterdam*, p. 38.
55. Gille, *Origines*, p. 12.

Colbert was forced to follow the traditional patterns. He had permanent commercial agents at The Hague and Amsterdam and he bought large amounts of cannon in Holland as well as in Sweden, Denmark, Hamburg and Biscay.[56]

Of all people however, Colbert was the one who found this arrangement least satisfactory. In a letter dated 1666 he wrote unambiguously that

> it is highly necessary to be careful in purchases. It is better to buy French rather than foreign products even if the French ones are a little less good and a little more expensive. There is a double advantage in doing so: the state does not lose liquid assets and is not impoverished and on the other hand the subjects of His Majesty make a living and develop their skills.[57]

In accordance with these principles, just about 1665 Colbert started a general plan for the development of the French armament industry. He had two reasons for being in favour of iron guns: (1) iron guns were less expensive than bronze ordnance and (2) France had plenty of iron ore while she had to depend on imports for copper and tin. This second reason would not have necessarily appealed to a free-trade zealot. Colbert was a zealot, but not a free-trader. The deterioration of France's relations with Holland gave him reasons to carry on his plan as soon as possible. This was conceived as an organic whole for the entire country. The location of the iron ores and the availability of water routes for the transportation of the finished products dictated the choice of the areas: Angoumois, Périgord and Nivernais to serve the arsenals of the western coast; Burgundy, Lyonnais and Dauphiné to serve the arsenals of the northern and southern coasts. A most serious bottleneck was that of the shortage of technicians and Colbert had to get them from abroad.[58] Dis-

56. Basset, *Historique*, p. 990.

57. Clément, *Lettres*, vol. 3, part 1, p. 76.

58. By that time Swedish technicians were deservedly famous and Colbert invited to France Abraham and Hubert Jr De Beche, nephews to the famous Guillaume De Beche. (See above p. 53). Swedish workers were also brought to France. This emigration was a source of preoccupation for the Swedish government which asked the '*Bergskollegium*' to investigate. The results of the inquiry were presented to the government in 1669. It appeared that a number of Swedish workers sailed from Nyköping believing that they were being brought to some other part of Sweden. But they were brought to

trusting small enterprise, he promoted the formation of large private concerns which he favoured through large orders, preferential treatment in the supplying of iron ore, and help in the recruitment of labour. As usual with him, he carried out his plan with meticulous care, indefatigable energy and stubborn will-power. He was determined to the borders of blind obstinacy. When many of the newly produced cannon blew up at a test, he wrote to the navy commissioner in Dunkerque:

I do not know anything about the quality or the provenance of the iron [guns] that blew up at Boulogne. But I know that whenever naval equipment is manufactured in this Kingdom for the first time, our people always find it bad. Often it is bad ... but if the iron has been badly manufactured one should not find this surprising because it is difficult to have things done well at the first trial. If one keeps supplying the manufacturers with good samples and helps them in correcting their faults, one will eventually obtain products as good as those that are manufactured in foreign countries.[59]

This he wrote in 1670. But the following year even his faith began to be shaken and he wrote in a discouraged mood '*je ne sais plus qu'en croire*'.[60] Too many guns blew up on being tested.

The final outcome of fifteen years of efforts was a mixture of some success and much failure. The factories established in Périgord and Angoumois gave satisfactory results. By 1680 they were manufacturing iron guns that were 'certainly lighter and better than those of Sweden' and could be produced in 'quantity sufficient to satisfy the needs of the ports of the Western coast'.[61] But the manufactures established in Nivernais and Burgundy turned out to be a complete failure and admittedly it was there that the greatest efforts had been spent.

The reason for Colbert's very modest results lies essentially in the fact that he was too much in advance of his times. Chemistry had not yet discovered the negative role of sulphur and the positive role of phosphorus in the casting of iron. There was no way for the technicians of Colbert's time to know that the iron

---

Lübeck and from there to Hamburg and finally to France. A few escaped and one of them, Anders Sigfersson, returned to Sweden in 1675.

59. Clément, *Lettres*, vol. 3, part 1, p. 76.
60. ibid., p. 379.
61. ibid., part 2, p. 379.

ore of Périgord was fitted for the contemporary methods of casting while the iron ore of Nivernais was not. The brittleness of the cannon cast in one place and the reliability of the cannon cast in another remained an unsolved mystery. On the economic and social level, Colbert had intelligent, devoted and very active collaborators, but the country as a whole did not follow him. The aristocracy was lagging behind the times. The 'Third Estate' was moving slowly and along overgrown paths. Already before the death of the great Minister (1683), all that remained of his industrial construction showed traces of weakness and fatigue. The situation grew progressively worse until the 1730s and it was not until the second half of the eighteenth century that France was able to reverse the trend and build an important armament industry.

The European situation in regard to the production of guns about the middle of the seventeenth century was vastly different from that prevailing two centuries before. The geographical distribution of the industry had radically changed. More important than that, European productive potential had increased enormously, and Europe had become much more formidable. The most relevant fact was the appearance of effective iron guns. This gave Europe the possibility of expanding her artillery park at a relatively low cost, while progress in technology and business organization allowed more efficient use of available resources.

Any effort to evaluate total European production is bound to be subject to an exceedingly large margin of error, but some estimates, no matter how approximate, may at least serve to give some idea about orders of magnitude. About 1650 Sweden was able to produce probably 1,500 to 2,000 metric tons of cast-iron cannon per year.[62] England produced less than 1,000 tons.[63]

62. Table 1, above on p. 55, shows that just after the middle of the seventeenth century Sweden could easily export more than 1,000 tons of cast-iron cannon per year.

In 1655 the Swedish Crown established a royal monopsony for all cast-iron cannon produced in Sweden. The management was entrusted to B. O. Cronberg who was in charge until 1662. From his accounts (*Rechnung über Eiserne Stücken* ... 1655–63; in Kammararkivet, Stockholm) it appears that between June 1655 and November 1662, 10,135 cannon for a total weight of 10,253 tons (= 75,390 skepps) were sold to the Crown or exported

These were by far the most important areas of production. The other main areas were Biscay in Spain, Western Germany, the Tula region in Russia, and Périgord in France. We possess some quantitative information, only for the Tula region which was able to produce possibly 250 to 300 tons per year.[64] Anyhow it is unlikely that the total production of all these areas exceeded the total combined production of England and Sweden. If this was the case, we may calculate a total maximum productive capacity potential of about 5,000 tons per year of cast-iron cannon. To this the production of bronze cannon should be added. Considering that in the biggest European navies iron guns had become predominant after 1650 and that iron guns were largely used also on land (although mostly in fortresses) one may venture to say that the total European production of bronze cannon can hardly have exceeded 5,000 tons per year. But this is pure speculation. We are on firmer grounds when we try to assess the qualitative gains made in European armament production. By the early sixteenth century naval bronze guns had attained so high a state of perfection that they remained substantially unaltered throughout the next century and a half. By 1650 naval iron guns were not as good as bronze ordnance, but had reached a satisfactory degree of efficiency. Field artillery was the weak point of European munitions and it was in this sector that revolutionary changes occurred in the course of the seventeenth century.

At the beginning of the seventeenth century the so-called leather guns appeared. Romantic writers have created around

---

under licence. This would give an average of about 1,370 tons per year

In November 1662 the administration of the Crown's monopsony was taken over by the *Generalfaktorikontoret* headed by Abraham van Eijck and Johan von Friesendorff. According to a statement by van Eijck (cf. *Svensk Biografiskt Lexicon* ad vocem) between November 1662 and November 1666 the *Generalfaktorikontoret* sold 9,457 tons of iron cannon (= 69,534 skepps), i.e. a yearly average of about 2,360 tons.

The greater part of the output came from the factories at Finspong, Nävekvarn, Åckers, Bränn-Ekeby, Fada, Huseby, Svärta, Julita, and, in the second part of the century, also from Stavsjö, Hällefors and Ehrendal.

63. English output of guns *and* ammunition has been estimated all together at 800–1,000 tons around 1600 (cf. above p. 65). After that time the charcoal crisis must have adversely affected the levels of production.

64. Strumilin, *Istoriia*, pp. 104*ff*.

them a glorious legend, but in reality leather guns were neither a practical nor efficient weapon. In shooting power they were admittedly inferior to metal guns. Moreover they had an exceedingly short life as individuals. Because of these rather substantial defects, they also had a very short life as a species. But times were ripe for change. In April 1629 Stockholm's royal foundry produced the first 3-pound '*regementsstycke*'; a piece that, while weighing only 123 kg. and being therefore highly mobile, could fire three shots in the time that it took a musketeer to fire one. The long quest for effective field artillery, a quest that had begun in the fifteenth century, was therefore solved by European technicians in the first half of the seventeenth century. The immediate results were visible in inter-European wars. The successful campaigns of Gustavus Adolphus cannot be adequately understood if one does not take into due account the technological achievements of the Swedish foundries. But the new weapons soon opened a new chapter also in the history of European expansion. With the appearance of the *regementsstycke* the balance of power pointed to a further remarkable shift in favour of Europe.

In the descriptions of European expansion the references to European superiority in armament are generally given in static terms. The fact is, however, that after the first wave of expansion in the fifteenth century the European potential in armament production increased dramatically from a quantitative as well as from a qualitative point of view. This made any adjustment of the non-European peoples extremely difficult and their defence problematic, especially because the European progress in the manufacture of guns was accompanied by an equally remarkable progress in the construction of men-of-war and by the development of new techniques of naval warfare.

It has been indicated in the previous paragraphs that until the middle of the seventeenth century the main shortcomings of artillery for land warfare were slow rate of fire and lack of mobility. The limitations in mobility, however, were overcome in naval warfare and this accounts for the early, extensive and successful adoption of cannon on board European vessels. Guns made in Tournai were aboard the ships that Louis de Male sent to attack Antwerp in 1336.[65] Genoese galleys had on board firearms in 1338 and Venetian boats carried bombards in 1380.[66] Guns were possibly aboard Spanish ships in 1359 and 1372 and by 1381 Catalan merchantmen carried artillery.[67]

These developments came to coincide with a complex set of circumstances which have been often described[68] and therefore I will cite here only briefly: the closer contacts between Mediterranean and Northern navigation; the use of compass and the development of open sea navigation in the Atlantic area; the shortage of labour due to recurrent plagues after the middle of the fourteenth century and the improvement in the standards of living of the masses which made it more difficult to recruit oarsmen for the galleys; the expansion of trade in the course of the fifteenth century. It is difficult – not to say impossible – to assess the relative importance of each one of these circumstances, but there is no doubt that as a whole they exerted a remarkable impact on the development of shipbuilding and especially on the development of the sail boat. By the end of the fifteenth century the sailing ship had developed to such a point that, while its 'rig would have appeared utterly strange to a navigator of the earlier age ... it seems no exaggeration to say that the ship captains of the age of the great geographical discoveries, or even those of a generation before, would have had but little to learn before taking charge of a ship of Nelson's day'.[69]

This progress, no matter how substantial, was of course purely empirical and often fortuitous. It was unequal, unsystematic, full of successful experiments as well as unconsequential trials. Oversimplifying the matter we may say that the main aspects of the progress were: the adoption after 1300 by Mediterranean merchantmen of the square-rig on the main mast and the consequent

65. Henrard, *Documents*, p. 240.

66. In October 1338 a French fleet appeared in Southampton water. Some of the galleys were from Genoa, manned by Genoese in the service of the French King. They had on board *pot de fer* with gunpowder and forty-eight iron bolts presumably to be used for breaking in the gates of the town (Ruddock, *Italian merchants*, p. 32). In 1380, according to Genoese chroniclers, the Venetians used bombards on some of their vessels (Montù, *Artiglieria*, vol. 1, pp. 119–21).

67. De Artiñano, *Arquitectura naval*, pp. 43–4.

68. Cf, Parry, *Age of Reconaissance*, chapt. 3.

69. Lane, *Venetian Ships*, p. 35.

transition from the one-masted to the three-masted ships; the noticeable increase in tonnage of merchantmen during the fifteenth century; the greater reliance on artillery for attack and defence.

So much has been written on this subject that I find unnecessary to repeat what is already well known. But I want to emphasize that the above mentioned developments were the result of effective interplay between Mediterranean and Northern ideas of build and rig: the caravel as well as the carrack were neither essentially Northern nor Southern in type. From 1250 to 1450, the differences between the North and the South had progressively levelled out. But after the middle of the fifteenth century a new and quite important difference developed between Atlantic and Mediterranean naval powers, namely a difference in attitudes toward the type of ship to use in warfare.

A long tradition going back to Roman times had established a kind of 'division of labour' in Mediterranean waters between the 'long ship' equipped with oars (the galley) and the 'round ship' dependent on sail, the former being mainly used as warship and the latter as merchantman. This 'division of labour' was shaken at the end of the thirteenth century by the appearance of the 'great galley' which was a hybrid vessel 'designed to combine not only some of the advantages of oarship and sailing vessel but also those of warship and merchantman'.[70] Further shaking occurred in the course of the fifteenth century when well-gunned round ships were used by Venice as auxiliaries to her armadas or to patrol the seas and hunt the ever-present pirates. However, up to the end of the seventeenth century the galleys, and the galleys alone, remained the backbone of the Mediterranean war fleets. This was true for Venice as well as for Genoa, Turkey and the Holy Order of Malta.

Serviceable enough in the Mediterranean however, the galley could not stand up to the long rollers and the fierce sou'westerlies of the Atlantic. As the potentialities of the round ship were gradually developed in the course of the fifteenth century, Atlantic naval powers turned to the sailing vessel and made of it the core of their fighting navies. It has been recently suggested that 'it was the English who upset the convenient division of labour'

70. Lane, *Venetian Ships*, p. 24.

between galley and sailing ship and made of the latter a successful naval fighter.[71] It is difficult, however, to say who was the first: most probably there was a series of multiple and interdependent influences among Atlantic nations. Henry VII of England (1485–1509) had especially built for his war-navy in 1487 two sailing ships manned with guns, the *Regent* and the *Soveraign*, which represent a good and early example of the new developments that were taking place in the Atlantic area. Garcia de Resende (1470–1536) reports that in Portugal King John II (1481–95)

> used to spend much money in building great vessels manned with guns. But, as he was ingenious in every task and knew very much in matter of artillery, in order to protect the coast with more security and less expense, after much experimentation he discovered and directed how many a great cannon might travel in small caravels and fire so low that their shot went skimming over the water. He was the first who made this invention. A few of such caravels could force many big ships to surrender because they were heavily gunned, but at the same time they were also small and highly manoeuvrable so that the big vessels could hardly hit them; and for a long time the caravels of Portugal were very much feared on the seas.[72]

Exchanging oarsmen for sails and warriors for guns meant essentially the exchange of human energy for inanimate power. By turning whole-heartedly to the gun-carrying sailing ship the Atlantic peoples broke down the bottleneck inherent in the use of human energy and harnessed, to their advantage, far larger quantities of power. It was then that European sails appeared aggressively on the most distant seas.

Aboard the sailing ship, guns were first placed on the decks of the castles. Later on, as the guns became heavier, the larger ones were carried on the upper deck firing over or through the bulwarks while the lighter pieces were carried inside – rather than above – the fore castle and the after castle. A caravel could, under special circumstances, carry up to 30–40 pieces, but ordinarily it did not carry more than 15. Bigger ships carried much more: Henry VII's *Soveraign* had 141 guns, but 110 were iron 'serpentines', that is light breech-loaders mostly mounted in the castles.

71. Lewis, *Armada*, pp. 64–5.
72. De Resende, *Chronica*, chapt. 181.

At the beginning of the sixteenth century an important innovation was introduced: the cutting of ports in the actual hull of the ships – as distinct from the superstructures – so that cannon could be mounted not only on the upper deck or on the castles but also on the main deck. Traditionally the date of the innovation is fixed at 1501 and the innovation itself is attributed to a Frenchman. The discovery was of very great importance: it gave the bigger ships the possibility of increasing their armament vastly. Mounting the guns on the main deck not only made it possible to mount many more, but it also made possible the use of much bigger pieces without imperilling the stability of the vessel. The English warship *Harry Grace à Dieu*, built in 1514 according to the new technique, carried no less than 186 guns among which there were two brass culverins of about 4,500 lb. each and a brass curtall of about 3,000 lb. all placed on the lower deck. The *Harry* was the crack ship of her day: she was launched in the presence of the Court, the ambassadors of both the Emperor and the Pope and a distinguished assembly of bishops and nobles. The 'demonstration effect' set immediately to work. The *Harry Grace à Dieu* was followed in 1527 by the French *Grand François*, in 1534 by the Portuguese *Sao João* which is said to have carried no fewer than 366 guns, in 1554–9 by the Swedish *Elefanten* with 71 cannon of which 24 were of bronze.

What these monsters looked like, we learn from the paintings of the time. With their colourful flags and their fantastically ornate castles they diffused an aura of grandeur and pageantry. They were magnificent as well as formidable but they were very clumsy to manoeuvre. In order to take the greatest possible advantage of the invention of the gunports and make the maximum use of the effect of the broadside, the pendulum had swung to the opposite of the '*muito ligeiras e pequenas caravelas de Portugal*'. The great ship had become a sea-fortress rather than a manoeuvrable weapon of war: 'a sluggish mover, clumsy to steer, very high in free-board and cumbered with flimsy castles in bow and stern'.[73]

Inevitably the pendulum had to move back. Shipbuilders set themselves to improve upon the manoeuvrability of the sailing ship without affecting fire-power and after 1550 their efforts

73. Lewis, *Armada*, p. 25.

resulted in the legendary galleon, a boat formidable in armament and swift in manoeuvring that could serve at the same time as a deadly warship and as an efficient merchantman. In relation to the previous type of great ship, the galleon was longer for its beam and with a hull built to some extent on galley lines; it was lower in freeboard and much less 'castled' especially in the bow (not to speak of improvements in the deck arrangements). Its name (which originally covered quite different types of vessels), and its shape, indicate that its genesis was influenced by the design of the dexterous galleys and this fact in its turn can reasonably indicate a Spanish origin. Yet, there is no doubt that those who adopted and perfected the new type of vessel most quickly and got the best out of it were eventually the English and the Dutch.

The Mediterranean people lagged behind. If there were some who advocated the adoption of the new types of vessels, there were too many who stuck to the glorious old traditions, who emphasized the advantages of the galley without recognizing its disadvantages, who thought that the effective destruction of the enemy could only be obtained by the time-tested techniques of ramming and boarding. Also, the physical conditions of the environment were more favourable to the traditionalist's views in the Mediterranean than in the Atlantic. When the Northern galleons began to bother the Venetian navigation in the Mediterranean, lengthy discussions took place in Venice on how to cope with the molesting intruders. The traditionalists who favoured the use of the galleys as the main fighting force turned out to be the vast majority. After an early experiment with sailing warships in the first decades of the sixteenth century, this type of vessel was discarded for naval action until the beginning of the following century. A '*galeone*' was eventually produced in 1608, but it turned out to be a clumsy 'floating fortress'. There was no tradition of galleon building in the Venetian arsenal nor were the crews sufficiently expert in the manoeuvring of the big men-of-war. Incapable of accomplishing the transition to the new ways of naval warfare, Venice had to call on England and Holland in order to resist Spain in 1616–19 and the powerful English and Dutch galleons came to the Mediterranean to protect what in the previous centuries had been the greatest sea-power of Europe.[74]

Spain herself was half Atlantic and half Mediterranean, and the Mediterranean tradition weighed heavily on her development as a sea power. It has been written that the defeat of the Armada in 1588 was due to the failure of the Spaniards to discard Mediterranean conceptions of naval warfare.[75] The thesis is too extreme, because institutional arrangements and the attitude of the crown bear their good share of responsibilities, but it does contain more than a grain of truth.

The Spaniards kept overloading their ships with soldiers for boarding tactics and proved incapable of dismissing altogether the oar-propelled galleys.[76] The Italian experts kept claiming unreasonably that 'to hit the enemy at long distance with artillery cannot be the purpose of a navy, the main object being that of ramming and boarding'.[77] The English Commission of Reform of 1618 declared instead: 'Experience teacheth how sea-fights in these days come seldome to boarding or to great execution of bows, arrows, small shot and the swords, but are chiefly performed by the great artillery breaking down masts, yards, tearing, raking and bilging the ships, wherein the great advantage of His Majesty's navy must carefully be maintained by appointing such a proportion of ordnance to each ship as the vessel will bear.'[78] Unencumbered by the outdated traditions of the Mediterranean type, limited in the availability of manpower, incorrigibly addicted to privateering, the English came to rely exclusively on manoeuvring by the wind and the effectiveness of the broadside. Few in numbers but aggressive and tenacious, scarcely imaginative but receptive and skilful, they produced, as Botero said, 'vessels very

74. Cf. Tenenti, *Venezia*, pp. 174–86.

75. Lewis, *Armada*, pp. 61–80.

76. When the Spaniards decided to invade England, they planned to include in the Armada 40 galleys and 6 galleasses. In 1588 the original plan was changed and the Armada sailed from Spain with 4 galleys and 4 galleasses. England at that time had only one galley 'which she wisely kept out of harm's way in the Thames' (cf. Lewis, *Armada*, pp. 59 and 62).

77. Gentilini, *Bombardiere*, pp. 26–7. According to Italian experts such as Da Canal and Busca the main purpose of naval artillery was that of hitting and disorganizing the enemy just before boarding the enemy's vessel. Guns should therefore be discharged at closed distance. Cf. Tenenti, *Cristoforo da Canal*, p. 38.

87. Quoted by Robertson, *Naval armament*, p. 21.

light and very well gunned'[79] that 'in a thousand ways bothered the huge ships of Spain'.[80] The Dutch followed the same line.[81] Within Europe herself, supremacy was gained by those nations which shifted more completely to guns and sails. The era of human energy was over and the era of the machine was beginning to open up.

79. Botero, *Relationi*, part 2, book 1, p. 257.

80. Botero, *Aggiunte*, p. 67.

81. The Dutch were able to put to an end the Portuguese maritime predominance in the East owing not only to a much larger navy but also to better ships. At the middle of the seventeenth century, Father Vincenzo Maria (*India Orientale*, p. 458) noticed that 'the [Portuguese] galleons are noteworthy because of their great size and their many facilities. Each one looks like a castle and is furnished with 80 or more bronze guns. The deck is so spacious, that the sailors often play ball. The rooms are numerous, spacious and with plenty of headroom so that the galleons resemble comfortable houses rather than vessels. The ropes are handled for the greatest part with the aid of capstans. The planking is thick enough to resist gunshots. In short, these vessels would be unequalled if they were not such sluggish movers and could be better manned. The Dutch vessels which are handier to manoeuvre by the wind, overcome the Portuguese galleons very easily. The Dutch can take to flight when the wind is favourable to the enemy and attack when the enemy is handicapped by low wind. To the Dutch any small wind is enough, while for the Portuguese vessels a half-gale is necessary for movement.'

Chapter 2

Guns and Sails Overseas

Thanks to their vicinity to Europe, the Moslems became acquainted with Western artillery when this was still in its infancy and they were quick to borrow it. In 1331 the Moorish King of Granada Mohammed IV attacked Orihuela and Alicante and allegedly used artillery.[1] From Spain the new technique moved to North Africa and the Middle East. Cannon was used in these areas by the Mamluks possibly in the 1350s,[2] certainly by the 1360s.[3] The Ottomans manufactured cannon in Asia Minor in 1364 and used guns in 1387 against the Karamans and in 1389 in the plain of Kossovo against the coalition army of Serbia, Bosnia, Herzegovina and Albania.[4]

As has been indicated in the previous chapter, fifteenth-century artillery was good only for battering walls, but it was precisely this quality that made it attractive to the Turks. Moslem armies had always enjoyed an indisputable superiority over their Western adversaries in the open field. Their advantage resided in the larger size of their forces, better discipline and superior tactics based on the high mobility of their light cavalry. The only efficient obstacle that the Europeans could oppose to the Turkish forces were

1. Cf. Fernandez Duro, *Disquisiciones*, vol. 1, p. 18; some authors however (e.g. Sarton, *Introduction*, vol. 3, p. 725) maintain that the evidence is not conclusive.

2. The fact is stated by later chroniclers, Ibn Buhtur and Ibn Iyas. Cf. Ayalon, *Mamluk*, p. 3.

3. According to two reliable eye-witnesses, the encyclopedist Al-Qualquashandi (*Subh al-a' shā*, vol. 2, p. 144) and the historian Ibn Khaldun (*Kitab al-ibar*, vol. 5, p.456) cannon were used in Alexandria and Cairo in the years 1366–8.

4. Daniemend, *Omanli Tarihi*, vol. I, p. 73; Uzunçarsili, *Kapukulu Ocaklaji*, vol. II, p. 35 *ff*.

defensive bastions. Against the walls of fortresses and towns, the dreaded Moslem horsemen were powerless and when faced with good battlements the Turks had to resort to long sieges for which they were organizationally unprepared. In artillery they found the kind of weapon they most badly needed.

It is truly remarkable how quickly the Moslems learned the new technique. But no less remarkable is the fact that they never succeeded in going beyond the initial stage. They never developed artillery into a field weapon. Occasionally the Mamluks used light guns mounted on camels and the Ottomans used guns at Kossovo and at Mohacs. But one perceives a distinct reluctance on the part of the Moslems to make use of artillery in battles of movement: a reluctance that developed into sheer inability. There were various reasons for this. To begin with, the Moslems did not feel a strong need for supplementary new weapons in open fighting. They were perfectly conscious of their tactical and strategic superiority and on the other hand early artillery was conspicuous for inefficiency in the field. Furthermore, ingrained traditions and social structures stood in the way of an extensive adoption of field artillery. As Dr Ayalon has brilliantly shown, 'horsemanship and all it stood for were the pivot round which the whole way of life of the Mamluk upper class revolved and from which it derived its courtly pride and feeling of superiority'. No Mamluk knight would have ever agreed to be deprived of his horse and to be reduced to the humiliating status of a foot soldier. On the other hand, as long as the social structure of the kingdom remained essentially feudal, the backbone of the army was necessarily represented by the knights and there was no possibility of giving a major role to other corps. Artillery was left to the black slaves who were the most despised human element in the kingdom: an unfortunate lot of people whose only chance of military advancement was by being castrated and incorporated into the corps of eunuchs.[5]

The Ottomans were slightly more flexible, for one of their main corps, the Janissaries, had been mainly composed of infantry archers before the introduction of fire-arms. The Janissaries were very early equipped with hand-guns and in all likeli-

5. For all that precedes cf. Ayalon, *Mamluk*, chapt. 3 and especially pp. 61–71.

hood this was one of the main reasons for the final victory of the Ottomans over the Mamluks in the first decades of the sixteenth century. But also in the case of the Ottomans, a traditional taste for the mêlée and for horsemanship, and the social predominance of the mounted warrior, acted as powerful obstacles to the adoption of field artillery. Essentially the Ottomans made good use of guns only in siege operations.

There was something primeval and instinctive in the hatred with which these warriors whose not too distant ancestors lived as nomads in the steppes, used guns against the walls of their enemy's urban settlements. In their primitive fury they mostly strove for huge guns that hurled enormous stone balls. The story of their conquest of Constantinople is very well known but it is of such significance that it is certainly worth retelling. The walls of Constantinople, even in their days of decay under the Paleologi, were still the most formidable system of defences in Europe. These walls gave nightmares to Mohammed when he took in hand the reduction of the fabulous town: the Turkish chroniclers tell us of the nights in his tent when the obsessed Mohammed, incapable of rest, kept drawing plans and inventing schemes to destroy the sturdy defences of the besieged town. Only artillery could help him out, and of this he was aware. It so happened that a skilled Hungarian gun-founder named Orban, who was in the service of the Greek Emperor, was dissatisfied with the pay he got from the Byzantines, and came to the Turkish camp offering his skills for better remuneration. Mohammed welcomed the renegade most generously, and Orban produced a gun of extraordinarily great size. The piece was tested against an unlucky Venetian galley that happened to pass in the straits: one huge stone ball hurled from the gun was enough to crack the ship and sink it. Mohammed exulted and in his excitement ordered Orban to produce a second gun twice as big. So Orban produced '*Mahometta*', a huge gun of fantastic size, made of hooped iron, which could throw stone balls of about one thousand pounds weight. It took from 60 to 140 oxen to draw the gun, more than one hundred men to manoeuvre it, and two hours to reload it. The noise it made when shooting – chroniclers say – caused miscarriages among pregnant women. It was by far the greatest gun ever built. But it was a failure. It cracked on the second day of the siege and went permanently out of order on the fourth or the fifth day. The guns of lesser calibre, however, successfully carried on their task; the town was taken and days of unrestrained savagery, unimaginable cruelty, unspeakable horror followed.

There are two points in this story which are especially worth emphasizing. The first relates to the origin of the renegade gun-founder. According to some chroniclers, he was a Hungarian, therefore a 'Westerner'. It is difficult to say how reliable the tradition is, but even if it were proven false, its very existence is significant. Although the Moslems had gun-founders of their own, whenever the possibility arose, they always tried to get Western technicians: '*les Turcs les retiennent quand ils les peuvent attraper*', wrote a seventeenth-century French economist.[6] In many a case, they did not have to 'capture' the needed craftsmen. Then, as always, there were many 'technicians' ready to sell their skills – to the devil, if the price were high enough and Orban, if such a man ever existed, was no exception. The eagerness that the Turks showed in obtaining Western gun-founders clearly indicates that although they adopted Western techniques very early, they never succeeded in going beyond the stage of 'follower country'. The West kept moving ahead, and especially after the middle of the fifteenth century it moved progressively faster. This brings us to the second relevant point of the story, namely the obsession of the Turks with huge guns.

To be sure, when confronted with the descriptions that the contemporaries offer of the huge Turkish guns, it is difficult to escape those feelings of scepticism that Voltaire sharply expressed in the '*Essai sur les moeurs*'.[7] However the concrete evidence that we possess seems to support the descriptions by the old chroniclers rather than the criticism by Voltaire. In 1867 the Sultan Abdul Aziz sent as a present to Queen Victoria a fifteenth-century Turkish cannon. It is not one of the biggest that the Turks ever possessed; nevertheless its calibre measures 25 inches, the barrel casing is 5½ inches thick and the monster as a whole weighs no less than 18 tons and 7 hundredweight: to this day it can be admired (and measured) at the Tower of London. In 1537 at Diu the Portuguese captured a gigantic bronze gun. The piece can

6. De Montchrétien, *Traicté*, p. 51.
7. Voltiare, *Moeurs*, vol. 12, pp. 100–101.

still be seen as one of the show-pieces of the Military Museum at Lisbon. It weighs more than 19 tons and it could fire a shot of 110 lb. Botero certainly indulged in exaggeration when he wrote early in the seventeenth century that 'the Turks have such huge guns that they can tear down battlements only by their noise,'[8] but behind the hyperbole there was an undeniable reality.

For the manufacture of the huge Turkish guns, much copper was needed, but the Ottomans possessed rich mines in Anatolia and plenty of slaves to operate them. Serious difficulties hindered the transportation of the big guns, but the Sultans resorted to the drastic solution of casting the pieces on the very place of the operations. In their obsession, they never admitted the short-comings of huge cannon, nor did they realize that it was becoming obsolete. As late as the eighteenth century, Baron de Tott related this humorous episode of the Turkish–Russian war (1768–74):

On the castle which commanded the Strait, the Turks had placed an enormous piece of ordnance which would carry a marble ball of eleven hundred pounds weight. This piece, cast in brass, in the reign of Amurath, was composed of two parts, screwed together. . . . I could not make use of this enormous cannon in the outworks, and the Turks murmured at my paying so little regard to a piece of artillery which, no doubt, had not its equal in the Universe. The Pacha made some remonstrances to me, on that head. He agreed with me that the difficulty of charging it would not allow, in case of an attack, firing it more than once, but he urged, this single discharge would be so destructive and reach so far, that it would be, alone, sufficient to destroy the whole fleet of the enemy. It was easier for me to give way to this prejudice than overthrow it, and, without changing my plan of defence, I could, by cutting through the epaulement in the direction of this piece, allow it room to be fired, but I was willing first to judge of its effect.

The crowd about me trembled at this proposal, and the oldest among them asserted there was a tradition that this piece, which had never yet been discharged, would occasion such a shock as must overturn the castle and the city. It was indeed possible it might shake some stones out of the wall, but I assured them they would not be regretted by the Gran Seigneur. . . . Never, certainly, had any cannon so formidable a reputation. Friends and enemies were alike to suffer from its fury. To load this piece of artillery it required no less than three hundred and thirty pounds weight of powder, and I sent to the Head Engineer to prepare a priming. All who heard me give this order

8. Botero, *Relationi*, part 2, book 4, p. 339.

immediately disappeared, to avoid the predicted danger. The Pacha himself was about to retreat, and it was with the utmost difficulty I persuaded him that he ran no risk, in a small kiosk, near the corner of the castle, from whence he might observe the effects of the ball.

Having succeeded in this, nothing remained but to inspire the engineer with courage, who, though he was the only one who had not fled, shewed no great resolution in the remonstrances he made to excite my pity. I, at last, rather silenced than animated him, by promising to expose myself to the same danger. I took my station on the stone-work, behind the cannon, and felt a shock like that of earthquake. At the distance of three hundred fathoms I saw the ball divide into three pieces, and these fragments of a rock crossed the straits, and rebounded from the water to the opposite mountain . . .[9]

The episode related by Baron de Tott was certainly not the last of its type in the fantastic story of the Turkish artillery. As late as 1807, when Sir John Duckworth's squadron forced the Dardanelles, the English sailors, to their incredulous astonishment, saw many a stone ball of enormous size being noisily hurled against their vessels.

Until the middle of the fifteenth century, also, Western gunfounders, striving for super-weapons, had dreamed of monstrous guns, but at the very moment when Mohammed II was firing '*Mahometta*' the technicians in the West reversed the prevailing trend and devoted great efforts to the production of light field artillery. The Turks altogether failed to recognize the importance of the innovation and did not keep up with the new developments. Their undisputed superiority in battles of movement and the feudal structure of their army contributed to this failure. Also, with the sixteenth century, the Ottomans entered a period of social unrest and economic difficulties that hampered change and adaptation. Whatever the cause, it is certain that they continued to devote their main efforts to the production of siege ordnance[10] and lagged behind the West both in the production and the use of field artillery. As long as this type of cannon was in its infancy in Western Europe, the lag did not noticeably affect the balance

9. De Tott, *Memoirs*, vol. 2, part 3, pp. 66–9.

10. The Turks acquired special expertise in the use of siege artillery. They developed the technique of using composite batteries in which huge guns operated together with medium calibre pieces in demolishing walls. This technique is described by Collado, *Platica*, p. 13.

of power. But in the course of the seventeenth century and especially after the appearance of the Swedish *regementsstycke* European progress in the manufacture of highly mobile field-artillery with a good rate of fire became rapid and substantial. In the second half of the century, Raimondo Montecuccoli, the general who routed the Turkish army at the battle of St Gothard (August 1664) wrote on the basis of his own experiences:

This enormous artillery [of the Turks] produces great damage when it hits, but it is awkward to move and it requires too much time to reload and sight. Furthermore it consumes a great amount of powder besides cracking and breaking the wheels and the carriages and even the ramparts on which it is placed.... Our artillery is more handy to move and more efficient and here resides our advantage over the cannon of the Turks.[11]

With an artillery that was, as Baron de Tott saw it, 'formidable in appearance from the width of its bore, but little to be dreaded after the first discharge from the slowness with which these enormous pieces must be served',[12] the Turks were hopelessly doomed.

On land the balance of power between the Turks and the Europeans turned definitely against the Turks in the course of the seventeenth century. On the seas, it had turned against them more than a century earlier.

On 7 October 1571 at Lepanto a Christian fleet of 208 galleys, including six immense galleasses, engaged a Turkish fleet of 250 galleys. After a ferocious fight that lasted three hours, 80 Turkish galleys were sunk, 130 were captured and only 40 escaped from the wing. The West rejoiced and the whole Christian world joined in boasting of the importance of the great victory. The Pope exulted more than anybody else and in his excitement declared that by divine grace he had 'seen' the whole course of the battle while sitting in his chair in St Peter's at Rome.

The Turks did not appear very shocked. 'The infidels only singed my beard; it will grow again' said the Sultan reportedly and it does not seem that such a complacent reaction was a mere piece of propaganda. His main strength was on land and not on the sea. The Turks also knew that they had enough resources

11. Montecuccoli, *Aforismi*, p. 457.
12. De Tott, *Memoirs*, vol. 2, part 3, p. 38.

to rebuild their navy in a matter of months, as in fact they did.

Assessing the historical importance of the battle, Western historians (except when writing text books) generally tend to lean more toward the interpretation of the Sultan than toward that of the Pope. 'A great victory without consequences', they say, and blame the lack of unity in the Christian field, the rivalries between Venice and Spain and the involvement of Spain with England and Holland. However, Lepanto did not have 'great consequences' essentially because it was not a 'great victory'. Lepanto was an anachronistic battle, fought with galleys and a lot of boarding and ramming, at a time when novel types of vessels and weapons were opening a new era in naval warfare and showing the way to a new naval strategy. At Lepanto the victors were no less anachronistic than the vanquished: both parties were prisoners of outdated traditions and techniques. In the light of history, at Lepanto everybody lost.

Less glamorous and less publicized, the victories of the Portuguese against the Moslems in the Indian Ocean during the first half of the sixteenth century were historically of a much greater significance. Immediately after the arrival of Vasco da Gama at Calicut, the Moslems realized that the presence of Christian vessels in the Indian Ocean opened a new front, threatened their position and upset the traditional balance of power. In 1507 a fleet with no less than 15,000 men left Egypt under Admiral Mir Hussain aiming at the destruction of the Portuguese. But the Portuguese defeated it off Diu in 1509. The attack was renewed and new expeditions were launched, but despite isolated success, the Moslems failed in their efforts and the Christians established their sovereignty over the Ocean.

It has been suggested that the failure of the Moslems was mainly due to the lack of timber supplies. According to this view, 'the total absence of timber on the Red Sea and the Persian Gulf' greatly hampered the rulers of Egypt (before 1517) and of Turkey (after 1517) in their efforts to build a fleet for use in the Indian Ocean.[13] It cannot be denied that the lack of timber created problems for the Moslems, but this kind of difficulty they most certainly managed to overcome. In fact they armed one

13. Boxer, *Portuguese in the East*, p. 197 and Boxer, *Four Centuries*, p. 15.

fleet after another in the Red Sea in the course of the sixteenth century. The main reason for their failure lay rather in their outmoded techniques of naval warfare. The Turks as well as their traditional foes, the Venetians and the Holy Order of Malta, did not realize the importance and the implication of the naval revolution that the Atlantic powers had accomplished. They remained 'medieval' when the modern age had already begun. They used guns on their ships (although in their own primitive way) and they made use of sailing vessels. But essentially they remained heavily dependent on human energy: they stuck to the old tactic of ramming and boarding and the galleys were always the backbone of their fighting force.[14] Against the Portuguese on the Ocean they fought as they did at Lepanto. In 1551–2 when Piri's fleet moved to the Persian Gulf, in 1576–7 when a naval expedition was led by 'Aly Bey against Muscat, in the defence of Aden with Admiral Khidr Beg, it was always the large galleys, the *qādyrgas*, that carried the weight of the battle.[15] The Portuguese used galleys also but the backbone of their fleet was the large ocean-going gun-manned sailing vessel.[16] In a small closed sea the big galleys still had a few chances. It is not fortuitous that neither the Portuguese nor the Dutch ever succeeded in establishing themselves on the shores of the Red Sea. But on the ocean the galleys had no chances whatsoever. When they were not sunk by the guns of the great sailing ships they were easy prey to the fury of the elements.[17]

Late in the sixteenth century the Turks learned how to man ocean-going vessels. In the early decades of the seventeenth century the 'Moors' of the North African coast put together a considerable fleet of sailing vessels for privateering activities, but the Empire as a whole never made up for the late start. Western

14. Cf. the description of the Turkish warships given by Hāǧi Halifa and reproduced by Kahle (ed.) in Piri Re'is, *Bahrije*, pp. xxxiv-xxxviii.

15. Serjeant, *The Portuguese*, p. 179*n*, DD and p. 180*n*. EE. Cf. also the Ramūzī illustration reproduced in Kahle, *Piri*, p. xxxvii.

16. In his expedition to Aden (1517) Lopo Soares de Albergaria had 15 *naos* (large sailing vessels), 10 *navios* and *caravelas* (sailing ships smaller than the *naos*), 8 galleys, 1 *caravelao*, 1 *bargantin* and 1 Indian junk. In his expedition to Hermuz (1520) Diego Lopes de Sequeira had 11 *naos*, 2 galleons, 5 galleys, 4 square-rigged ships, 2 brigantines and 2 caravels (cf. Serjeant, *The Portuguese*, pp. 170–71, notes H and I).

17. As in 1554 when the fleet of Sīdī' Alī Ra'īs was destroyed by a storm.

naval technology kept moving ahead faster and the Turks lagged hopelessly behind – more and more so in the course of the centuries.

Having outflanked the Turks – Ottomans as well as Mamluks – the Europeans were confronted with peoples of vastly different technologies and civilizations: on the one hand the very primitive peoples of the Americas and Africa and on the other hand the highly developed and sophisticated peoples of Asia. The first, of course, had never dreamed of fire-arms, and we can paraphrase Paolo Giovio by saying that the noise of European artillery was enough to induce them 'into the worship of Jesus Christ'. As to the second, the picture is noticeably less simple.

Says an ancient Chinese text: 'Among the hills in the Western parts [of China] there exist beings in human shape, a foot or more in height, who are by nature very fearless. If offended they cause men to sicken with (alternate) heat and cold. They are called *Shan-sao*. By thrusting bamboos into fire, and producing a crackling sound the *Shan-sao* are frightened away.'[18] According to a tradition widely accepted among Chinese antiquarians, gunpowder and fire-works were developed in order to scare away the little demons with a noise that magnified the crepitating sound of the burning bamboos.[19] Precisely when this began, we do not know. There is no doubt, however, that as early as the tenth century A.D. the Chinese were using gunpowder and, by that time, not only to drive away the malignant *Shan-sao* but also in actions of war.[20] Whether the Chinese had invented the noisy compound or had borrowed it from Brahmin chemists is a question that does not need to detain us here.

Both in India and China, metallic guns, in the modern sense of the word, appeared long after gunpowder had begun to be used and after a long series of experiments with all kinds of rockets, missiles, incendiary projectiles, etc. Chinese bombards clearly dated 1356 and 1357 existed in Chinese museums and

18. Mayers, *Gunpowder*, p. 77.

19. Fire-crackers are known in China by the name of *p'ao-chu* which literally means 'crackling bamboo'.

20. Mayers, *Gunpowder*, pp. 85*ff.*; Wang Ling, *Gunpowder*, pp. 160–62; Needham, *Science*, vol. 1, pp. 131 and 134.

there is no doubt that weapons of this kind were used by the Chinese before the middle of the fourteenth century.[21] From China the knowledge of gunpowder and fire-arms spread to Korea, Japan, Java and other parts of Asia. For India, Ferishta relates that in the 1360s guns were largely used in the Deccan both by the Raja of Vijayanagar and by Muhammad Shah Bahmani. Muhammad Shah allegedly gave special attention to artillery. 'He founded a factory of fire-arms.... He had a separate department for artillery' and he employed '*Rumis*' (Turks) and '*Farangi*' (Europeans) well acquainted with the art of gunnery.[22] Ferishta wrote these lines more than two hundred years after the event, but he is generally a very careful and reliable historian and we have no reason to doubt his recital. We have previously noticed that by 1368 both Europeans and Turks were acquainted with artillery. Also the reference to Turkish gunners is not surprising because it is well known that Turkish influences were strong in those parts of India in the fourteenth century. Supplementary information comes from other sources. In Gujarat, guns were used in the battle of Malwa in 1421 and in 1457 in the bombardment of Mandalgash. In the second half of the century, Sultan Mahmud Beghra had, beside field and siege guns, artillery which was used in naval battles.[23]

Further research will undoubtedly add much needed detail to this fragmentary and incomplete recital. Much information is still badly needed on the problem of the direction in which the new technology travelled, its velocity and the agents of its propagation. But a few points seem reasonably clear. It is certain that artillery was known in Asia long before the arrival of the Portuguese. It is likely that Chinese guns were at least as good as Western guns, if not better, up to the beginning of the fifteenth century. However, in the course of the fifteenth century, European technology made noticeable progress and by 1498 'the armament of the Portuguese ships was something totally unexpected and new in the Indian (and China) seas and gave an immediate advantage to the Portuguese'.[24] European artillery was incomparably

21. Goodrich, *Note*, p. 211, Wang-Ling, *Gunpowder*, pp. 172–3 and Chow Wie, *Chung Kuk Ping Ji*, pp. 234–8.

22. Nadvi, *Use of cannon*, pp. 405–7.

23. Nadvi, cit., p. 407. Cf. also Crawfurd, *Dictionary*, p. 22.

24. Panikkar, *Asia*, p. 29.

more powerful than any kind of cannon ever made in Asia, and it is not difficult to find in contemporary texts echoes of the mixture of terror and surprise that arose at the appearance of European ordnance. The following passage from the *Rajavali* relates the first arrival of the Portuguese at Ceylon:

And now it came to pass that a ship from Portugal arrived at Colombo and information was brought to the King that there were in the harbour a race of very white and beautiful people who wear boots and hats of iron and never stop in any place. They eat a sort of white stone and drink blood. And if they get a fish they give two or three *ridé* in gold for it, and besides, they have guns with a noise like thunder and a ball from one of them, after traversing a league, will break a castle of marble.[25]

The Portuguese arrived in China in 1517, but the fame of their guns had preceded them since 1511 when they took over Malacca, or even before.[26] *Fo-lang-ki* was the name given to the frightening contrivances by those who reported about them: the name possibly meant 'Franks'[27] and the scholar Kou Ying-Siang had to warn his people that '*Fo-lang-ki* is the name of a country and not of a gun'.[28]

When the Portuguese fleet under Fernao Peres cast anchor off a quay of the port of Canton in 1517, the first thing they did was to fire a salute with cannon. In the words of Professor T. T. Chang 'it had never before occurred to the Chinese that in some part of the earth a demonstration of war implements could be also an expression of respect or courteous recognition.'[29] The populace was scared and the scholar-officials protested. The dreadful tales that they had heard about the 'barbarians with the big nose' and their formidable weapons were abundantly confirmed. Wrote censor Ho Ao a few years after the event: 'the *Fo-lang-ki* are most cruel and crafty. Their arms are superior to those of other foreigners. Some years ago they came suddenly to the city of Canton and the noise of their cannon shook the earth.'[30] And the scholar-official Wang-Hong reiterated: 'The *Fo-lang-ki*

25. Tennent, *Ceylon*, vol. 1, p. 418.
26. Pelliot, *Le Hōja*, pp. 204–7.
27. ibid., p. 204, *n*.244.
28. Mayers, *Gunpowder*, p. 96 and Pelliot, *Le Hōja*, p. 204, *n*.244.
29. Chang, *Trade*, p. 64.
30. Quoted by Chang, *Trade*, p. 51.

are extremely dangerous because of their artillery and their ships.... No weapon ever made since memorable antiquity is superior to their cannon.'[31]

The roar of European ordnance awoke Chinese, Indians and Japanese to the frightening reality of a strange, alien people that unexpectedly appeared along their coasts under the protection and with the menace of superior, formidable weapons and ruthlessly interfered with the natives' life. For many a responsible Asian, merchants apart, it was a nightmare. How to deal with the 'foreign devils'? To fight them or to ignore them? To copy and adopt their techniques and give up local habits and traditions or to sever all contacts with them and seek refuge in a dream of isolation? To be or not to be? The fog of Hamletic doubt began to pervade the soul of Asia – a doubt that was to plague Asia for centuries, a dilemma that was tragically unanswerable because both alternative solutions implied surrender and the only alternative to surrender was death.

Inevitably, an armament race developed. Cannon became a highly coveted commodity: a much sought after and well rewarded object of trade, the perfect gift for obtaining favours from a local ruler, the precious jewel of a princely dowry.[32] There was nothing that cannon could not buy – in fact as well as in fancy. In an old Javanese poem the beautiful princess Tarurôgô is sold to a Dutchman for three pieces of artillery.

Of course, the Europeans were not always willing to give away the weapons on which their supremacy rested. Often guns were turned over to the natives. In some cases the Europeans wanted to obtain from the local authority special privileges for their trade. In other cases it was a question of helping one potentate against the other according to the policy of 'divide and rule'.[33] Money-minded and motivated by profit, many a European

31. Pelliot, *Le Hōja*, p. 107, *n*.42.

32. 'The Kinge of Acheijn sent a peece of ordinance, such as for greatnes length and workmanship, the like is hardly to bee found in all Christendome, which hee gave in marriage with his daughter to the King of Ior, a town lying by Malacca, upon the coast of Sian.' (Linschoten, *Voyage*, vol. 1, pp. 109–10).

33. Around 1514–15 the Portuguese supplied the Persians with fire-arms and taught them the art of casting ordnance in order to increase the diffi-

was ready to sell anything to anybody: for pepper, good brass ordnance was sold to 'Babarautt, the arche pyratt'.[34] But it has to be admitted that the export of Western cannon to Asia never amounted to more than an insignificant portion of European cannon production.

On the other hand, the Asian potentates did not like to remain dependent on the Europeans for their armament. Whenever they could, they tried to set up gun foundries of their own and gun-founders were as eagerly sought after as cannon. In principle the Europeans were opposed to the idea of giving away their techniques. The Portuguese severely punished those who taught the natives the art of gunnery. To the Japanese officials who asked him to teach them gunnery, Pieter Nuyts, Dutch governor of Taiwan, replied: 'Japan rules the land with bow, arrow and sword, my country only with fire-arms, wherefore I cannot teach you this art.'[35] At Batavia, considering that the gun foundry was 'too close to the natives and Javanese, from whom this art should certainly be kept secret', the Dutch shifted the foundry to a less conspicuous place within the Castle walls.[36] In 1645, Pieter Antoniszoon Overtwater, chief of the Dutch factory in Nagasaki, wrote that 'these mortars are great jewels and one may well ask whether it had not been wiser never to have taught this proud and haughty nation about them . . . [as to the request by the Japanese authorities for the loan of a gun-founder] we ought to be rather evasive and fob them off with polite nothings.'[37] But in the course of time the Asian potentates always found some Europeans ready to teach them the art of making cannon. Ludovico Varthema reports that:

> Being then arrived in Calicut (1506) I found two Christians who were Milanese. One was called Ioan-Maria and the other Piero-Antonio, who had arrived from Portugal with the ships of the Portuguese and had come to purchase jewels on the part of the King. And when they

culties of the Turks, against whom the Persians were chronically at war (Godinho, *Repli vénitien*, p. 299). In 1541–2 the Portuguese gave cannon to the Abyssinians to help them against Ahmad Grāñ (Serjeant, *Portuguese*, p. 102).

34. Mundy, *Travels*, vol. 2, p. 316.
35. Boxer, *Jan Compagnie*, p. 28.
36. Boxer, op. cit., p. 27.
37. ibid., p. 38.

had arrived in Cochin, they fled to Calicut. Truly I never had greater pleasure than in seeing these two Christians. They and I went naked after the custom of the country. I asked them if they were Christians. Ioan-Maria answered: 'Yes, truly we are.' And then Piero-Antonio asked me if I was a Christian. I answered: 'Yes, God be praised.' Then he took me by hand, and led me into his house. And when we had arrived at the house, we began to embrace and kiss each other, and to weep. Truly, I could not speak like a Christian: it appeared as though my tongue were large and hampered, for I had been four years without speaking with Christians. The night following I remained with them, and neither of them nor I could eat or sleep for the great joy we had. You may imagine that we could have wished that that night might have lasted for a year, that we might talk together of various things, amongst which I asked them if they were friends of the King of Calicut. They replied that they were his chief men, and that they spoke with him every day. I asked them also what was their intention. They told me that they would willingly have returned to their country, but that they did not know by what way. I answered them: 'Return by the way you came.' They said that was not possible, because they had escaped from the Portuguese and that the King of Calicut had obliged them to make a great quantity of artillery against their will, and on this account they did not wish to return by that route. And they said that they expected the fleet of the King of Portugal very soon. I answered them, that if God granted me so much grace that I might be able to escape to Cananor when the fleet had arrived, I would so act that the captain of the Christians should pardon them, and I told them that it was not possible for them to escape by any other way, because it was known through many nations that they made artillery and many kings had wished to have them in their hands on account of their skill, and therefore it was not possible to escape in any other manner. And you must know that they had made between four and five hundred pieces of ordnance large and small, so that in short they had great fear of the Portuguese. And in truth there were reasons to be afraid, for not only did they make the artillery themselves, but they had also taught the Pagans to make it, and they told me, moreover, that they had taught fifteen servants of the King to fire *spingarde*. And during the time I was here, they gave to a Pagan the design and form of a mortar, which weighed one hundred and five *cantara* and was made of metal. There was also a Jew here who had built a very beautiful galley, and had made four mortars of iron. The said Jew, going to wash himself in a pond of water, was drowned. Let us return to the said Christians. God knows what I said to them, exhorting them not to commit such an act against Christians. Piero-Antonio wept incessantly, and Ioan-Maria said it was the same to him whether he die in Calicut or in Rome, and that God had ordained what was to be.[38]

Eventually, nostalgia and remorse won the case and the two gun-founders planned to escape, but they were discovered and lynched. This was the end of the story of Ioan-Maria and Piero-Antonio, but other European gun-founders were to come to Asia, as renegades or otherwise. In 1505 four Venetians arrived in Malabar to cast cannon.[39] In 1649 after years of 'polite nothings' the Dutch had to comply with the requests of the Japanese and the gunner Schaedel along with three other Dutchmen was sent to Yedo where they instructed the natives about guns and gunnery.[40] In China the Jesuits generously offered their services in exchange for the authorization to open Jesuit missions. Actually, the Chinese had cast *fo-lang-ki* at least since 1522: in that year they had got the help of two countrymen of theirs who were serving aboard Portuguese ships and were induced to desert by the clever and scheming Ho Jou.[41] The event was gloried over in the official imperial histories and the Imperial Ministry of the Interior 'recapitulating the previous merits of Ho Jou' eventually appointed him to the post of assistant-under-prefect in the under-prefecture of some under-division of the territory of Peking. However, it does not seem that left to themselves the Chinese could go very far in the production of cannon. The Jesuits made themselves very useful as intermediaries between the Imperial officials and the Portuguese authorities whenever the former wanted to buy Western guns in Macao. Most willing to oblige, they offered further services as instructors in the art of gunnery and gun-casting. To this job they were certainly well fitted. In the defence of Macao against the Dutch in 1622 it was the Italian Jesuit and mathematician Padre Giacomo da Rho who with a lucky cannon-ball struck the enemy's barrel of gunpowder and caused it to explode in the midst of the attacking party with devastating effects. Also in Macao, it was the Jesuits who showed great dexterity in gunnery when, angry at the Dominicans over a hot dispute, they took up their guns and aimed them at the

38. Varthema, *Travels*, pp. 260–62.
39. Whiteway, *Portuguese*, p. 37; Pieris, *Ceylon*, vol. 1, p. 445, *n*.18.
40. Boxer, *Jan Compagnie*, pp. 26–7 and 39.
41. Pelliot, *Le Hōja*, pp. 199-207.

monastery of St Dominic.[42] The Chinese could hardly have got better teachers. In the last days of the eighteenth century, an English diplomat travelling through China still heard about 'the two Jesuits Schall and Verbiest [who] took great pains to instruct them [the Chinese] in the method of casting cannon.'[43] As Dr Chiang puts it, 'while Buddha came to China on white elephants, Christ was borne on cannon balls.'

If some progress was achieved by the Chinese in the making and use of ordnance after their first contact with the Portuguese, it was remarkably slow. More than half a century after that first contact, Father Martin de Rada could still write that 'Chinese artillery (at least that which we saw and although we entered an armoury in Hocchin) is most inferior, for it consists only of small iron guns.'[44] Father Matteo Ricci held no different opinion: 'the arms supplied to the army are practically worthless for an offensive against an enemy as well as for self-defence.'[45] Also a Chinese text candidly admitted: 'the *Fo-lang-ki* use fire-arms with great skill. The Chinese on the contrary blow off their fingers or their hands or even their arms ...'[46]

By 1624 a Chinese military treatise boasted of some substantial progress and showed that the Chinese 'using their ingenuity improved upon the cannon of the Western barbarians and produced a gun larger in size and more flexible in use than the *fo-lang-ki*. The name of the gun is *fa-k'uang*.' According to the same text, the power of the *fa-k'uang* was such that a stone ball shot from it 'could pierce a wall, penetrate into a house, break a tree, turn human beings and animals into a mess of blood and also penetrate into a mountain for several feet'.[47] This huge gun was admittedly 'useful only in storming a fort or in trying to capture a strategic place'. Another military treatise indicates that guns of different calibres and dimension were being envisaged if not constructed, beside numerous other more or less fantastic

42. Boxer, *Fidalgos*, pp. 81 and 97.
43. Barrow, *Travels*, p. 302.
44. De Rada, *Relation*, p. 273.
45. D'Elia, *Fonti Ricciane*, vol. I, p. 104.
46. Pelliot, *Le Hōja*, p. 93, *n*.14.
47. Hu, *Ch'ou hai t'u pien*.

contrivances for vomiting fire.[48] But to a well-informed European writer at the middle of the seventeenth century, the guns of China still looked 'few and poor'. The efforts by Father Schall in the 1640s and by Father Verbiest in the 1670s did not substantially change the picture and China lagged hopelessly behind the West.[49]

The reasons for the Chinese failure to produce satisfactory artillery despite the technical assistance of the Jesuits, a good supply of raw materials, native wit and ingenuity, are not easily analysed. To ask why the Chinese did not produce good artillery is equivalent to asking why they did not industrialize, and to this kind of question there can hardly be an answer – or, rather, there are answers which are as vague and indeterminate as the question itself. But I would be inclined to remark that what was at issue was not only technical skill, but also taste, cultural pride and institutions. Ming Ch'ing China was 'a Confucian and phisiocratic State' where skilled artisans were not very numerous and did not have much status.[50] On the other hand, the demand was not such as to induce them to apply themselves to the adoption and development of Western technologies. The Imperial Court never developed that kind of enthusiasm for cannon that inspired the more technically-minded and more warlike monarchs of the West. Fearing internal bandits no less than foreign enemies and internal uprisings no less than foreign invasion, the Imperial Court did its best to limit both the spread of the knowledge of gunnery and the proliferation of artisans versed in the art. In the reign of Yung Loh fire-arms had already been obtained, but they were not permitted to be known publicly. By 1570 cannon had become well known to the populace and guns were placed at

48. Mao, *Wu-pei-chih*, especially vol. 50 but also vols. 51 to 56.

49. For the middle of the seventeenth century cf. Brusoni, *Osservazioni*, pp. 97–8 and Semedo, *Histoire*, p. 145. In the second half of the following century, Father Amiot gave a precise and detailed description (with measures and weights) of some of the cannon used by the Chinese troops: it appears that the few guns used were generally old and outdated, some dating back to 1636 (Amiot, *Art militaire*, pp. 375–84). At the end of the eighteenth century, according to De Guignes (*Voyages*, vol. 3, pp. 35–6) 'the greatest part of the guns that exist in China' were old pieces manufactured by Fathers Schall and Verbiest. De Guignes also noticed that the Chinese often used cannon balls made of dried mud. Cf. also Barrow, *Travels*, p. 302 and Keberg, *Ostindische Reise*, p. 101.

50. Ho, *Ladder of success*, pp. 41*ff.* and 56*ff.*

the gates of practically all walled towns. Still, the Imperial Court was very suspicious and did not like the idea of its subjects playing around with the art of gunnery.[51] The attitude of the Imperial Court was also heavily influenced by the fact that Chinese rulers had always been apprehensive of foreign influence, for they realized that the idea of the 'barbarians' being superior to Peking would be political dynamite.[52] If and when an Emperor adopted a more sympathetic attitude toward the Western 'barbarians' and their technology, conservative circles and officials busily manoeuvred to hinder the change. Sometimes it was just a question of feelings of envy and suspiciousness because of the repute acquired by some foreign 'barbarian' at the Imperial Court. But there was more than that. Among the gentry and the scholar-officials the prevailing cultural traits were not favourable to innovation. Of course it is not difficult to find examples of scholar-officials who recommended the adoption of the new weapons and the development of their manufacture. Wang Hong sent Portuguese guns to the Imperial Court in 1522 urging that they be used against the Mongols;[53] Hsü Kuang-chi strongly recommended the acquisition of guns in Macao in 1619 and again in 1630; Sun Yüan-hua wrote a report in 1621 recommending the adoption of Western cannon; Chü Shih-ssü managed to obtain Western guns through the intercession of missionaries and allegedly used them in the defence of Kuei-lin in 1648.[54] But the efforts of a few were not enough to compensate for the immobile conservatism of the many. The superior performance of Western science in matters of astronomy with all the implication that this had for the Chinese calendar and the Imperial administration, was too much of a shock to the scholar-officials to be accepted as a matter of course. The idea that the stylish poems and formal essays they learned were useless in the light of the engineering knowledge of the West, was more than they could bear. As the

51. On a few occasions the Imperial Government hired Portuguese guns as well as Portuguese gunners, but then it sent them home as quickly as possible. Cf. Boxer, *Expediçoẽs militares*; Pfister, *Notices*, p. 214, and Boxer, *Fidalgos*, p. 75.

52. Mu, *Hundred Flowers*, p. 76.

53. Pelliot, *Le Hōja*, pp. 199–201.

54. Hummel, *Eminent Chinese*, vol. 1, pp. 200, 317, 318 and vol. 2, pp. 686 and 912.

Chinese writer who conceals himself under the pseudonym of Mu Fu-Sheng cogently puts it:

Military defeat was the technical reason why Western knowledge should be acquired, but it was also the psychological reason why it should not be. Instinctively the Chinese preferred admitting military defeat, which could be reversed, to entering a psychological crisis: people could stand humiliation but not self-debasement. . . . The mandarins sensed the threat to Chinese civilization irrespective of the economic and political issues and they tried to resist this threat without regard to the economic and political dangers. In the past the Chinese had never had to give up their cultural pride: the foreign rulers always adopted the Chinese civilization. Hence there was nothing in their history to guide them through their modern crisis.[55]

Cultural pride stood tenaciously in the way of change. Traditional tastes and ingrained sets of value also opposed it. To the majority of the scholar-officials of Ming and early Ch'ing China nothing could be less attractive than noisy weapons and military innovations. Genteel initiates in an essentially humanistic culture, the officials of the Celestial Empire 'were amateurs in the fullest sense of the word, without interest in progress, leanings to science, sympathy for commerce or prejudice in favour of utility. Amateurs in government because their training was in art, they had an amateur bias in art itself for their profession was in government.'[56] Their culture being firmly established on tradition, they could not be persuaded – as Father Le Comte (1655–1728) noted – 'to make use of new instruments and leave their old ones without an especial order from the Emperor to that effect. They are more fond of the most defective piece of antiquity than of the most perfect of the modern, differing much in that from us [Europeans] who are in love with nothing but what is new.'[57]

Below the Imperial Court and the scholar-officials, cultural traits were not more favourable to change. The whole of Chinese society was imbued with a taste for tradition, individual virtuosity and colourful performance. Last but not least, both upper and lower classes shared an intense contempt for soldiers and military

55. Mu, *Hundred Flowers*, pp. 76–7.
56. Levenson, *Confucian China*, pp. 16–19.
57. Le Comte, *Empire of China*, p. 68.

matters. 'The military is one of the four conditions which are considered mean among them' wrote Father Ricci. And Semedo added: the weakness of their army is due: 1, to the long periods of peace they enjoyed; 2, to their high esteem for letters and their contempt for military matters; 3, to their way of appointing the army-captains, namely on the basis of (literary) examinations and not on the basis of military expertise.[58]

Nothing, I think, better than the following delightful episode can serve to illustrate the prevailing Chinese aura of patrician detachment and amateurish style. When in 1626 Yüan Ch'ung-huan had to defend Ning-yüan against the attacking Manchus and eventually decided to resort to 'foreign guns', the general direction of artillery operations was put in the hands of his Fukienese cook who, incidentally, put up a very good show.

If a Fukienese cook was good enough as captain-major of artillery against the 'barbarians' who were pressing against China from the steppes, much more was needed to fight against the 'barbarians' who were coming from the sea. But the scholar-officials of the Celestial Empire did not have much more at their disposal.

The disadvantage of the Chinese in matters of artillery in relation to the European powers was felt not so much on land as on the seas. As has been repeatedly said in the first chapter and will be emphasized again in the epilogue, until the middle of the seventeenth century the Europeans were unable to produce effective light field artillery. Their guns lacked mobility and in the open field their fire could be overwhelmed either by a massive assault or by manoeuvre. On the sea things looked quite different, and it was impossible for the Chinese or any other people to cope with the formidable European vessels.

It has been highly fashionable in the last few decades to sing the virtues of the Chinese junk, and there is no doubt that, as far as seaworthiness and navigational qualities in general go, these laudatory songs are deserved. Up to the end of the fourteenth century, the Chinese junk could compare honourably with Western sailing vessels. But the developments in build and rig that took place during the fifteenth century, made European

58. Semedo, *Histoire*, pp. 145–6.

vessels much more manoeuvrable than Chinese junks, especially in relation to manoeuvre by the wind. A Chinese text significantly remarked: 'The sails of the boats of the red haired barbarians are like spider's webs so that they can be turned in all directions for wind and go anywhere they want.'[59] Another problem was that the junk never developed into a man-of-war.[60] Like the Mediterranean galley, the Chinese war-junk remained essentially a vessel suited for ramming and boarding. With very high castles and no portholes for guns in the hull, the war-junk was fit only for the traditional way of fighting and such it remained.

Geronimo Roman, factor of the Philippines at Macao, wrote about 1584:

The King of China maintains a numerous fleet on this coast, although he is not at war with any one. In an island called Lintao, which is situated near this town [Macao], there is an arsenal, the director or *haytao* of which is continually occupied in superintending the building and equipment of vessels.... The admiral has the title of *Chunpin*. It is a very high rank, although inferior to the *Tutan*. He has a numerous guard and many drums and trumpets, which make a most agreeable music to the ears of the Chinese, but an insufferable din to ours.

[The Junks] have some small iron guns, but none of bronze; their powder is bad.... Their arquebuses are so badly made that the ball would not pierce an ordinary cuirass, especially as they do not know how to aim. Their arms are bamboo pikes, some pointed with iron, others hardened by fire, short and heavy scimitars, and cuirasses of iron and tin. Sometimes a hundred vessels are seen to surround a single corsair, those which are to windward throw out powdered lime to blind the enemy, and, as they are very numerous, it produces some effect. This is one of their principal warlike stratagems.[61]

The tone of the letter is definitely that of scorn but the matter of the substance is perfectly confirmed from other sources.[62] The Chinese threw arrows and fire-rockets other than lime powder, but it remains true that their naval tactics were and

59. Cf. Chang, *Trade*, p. 117.

60. Audemard, *Jonques*, p. 31; Worcester, *Junks*, vol. 2, pp. 348*ff.* and Din Ta San and Olesa Muñido, *El poder naval chino*.

61. Reproduced by Staunton (ed.) in the introduction to De Mendoza, *History*, pp. lxxix–lxxx. The 'stratagem ... of the powdered lime' was also used by the Indian pirates: cf. Vincenzo Maria, *Viaggio*, p. 245.

62. Cf. the report by Gaspar de Cruz reproduced in Boxer (ed.), *South China*, pp. 112–13.

continued to be based exclusively on ramming and boarding: 'All their use,' wrote Father Gaspar de Cruz, 'is to come many together, to compass the adversary ship and to board it.' As late as 1624 the military treatise *Ch'ou hai t'u pien* maintained that 'fire-arms can be used on large boats but the waves make aiming very hard. Chances of hitting the enemy are very slender. Even if one enemy boat should be hit, the enemy would not thus incur severe losses. The purpose of having fire-arms aboard is purely psychological, namely that of disheartening the enemy.'[63]

The Chinese, like the Turks and the Indians, lagged hopelessly behind the times in understanding the true potentialities of naval artillery and in learning the new naval tactics that artillery was mercilessly imposing. When they eventually realized that times had changed, it was too late. In the nineteenth century Feng Kuei-fen wrote:

> The most unparalleled anger which has ever existed since the creation of heaven and earth is exciting all who are conscious in their minds and have spirit in their blood: their hats are raised by their hair standing on end. This is because the largest country on the globe today, with a vast area of ten thousand *li*, is yet controlled by small barbarians. . . . Why are they small and yet strong? Why are we large and yet weak? . . . What we have to learn from the barbarians is only one thing, solid ships and effective guns.

In other parts of Asia, the diffusion of European technology encountered a less tenacious resistance. Unlike the Chinese, the Japanese never took their country as the centre of the world. Traditionally they looked beyond their border for foreign customs and techniques to copy and assimilate. For centuries their model had been China. When the Europeans arrived in Asia, the Japanese were not as hampered as the Chinese by self-centred cultural pride. 'The Japanese,' wrote Linschoten at the end of the sixteenth century, 'are sharpe witted and quickly learne anything they see,'[64] and Mendes Pinto observed: 'They are naturally addicted to war and in war they take more delight than any nation that we know.'[65] Under the leadership of their

63. Hu, *Ch'ou hai tu'pien*.

64. Linschoten, *Voyage*, vol. 1, p. 153.

65. Le Gentil, *Fernão Mendes Pinto*, pp. 151–8 and M. Collis, *The Grand Peregrination*, p. 148.

warlike barons, the Japanese set about copying Western fire-arms and succeeded in producing some cannon and great quantities of arquebuses.[66] In order to resist the Japanese invasion of the 1590s the Koreans too turned eagerly to the new weapons. They actually did better than the Japanese in the production of artillery although it seems that they lagged behind the Japanese in the production of hand-guns.[67] In India, Ceylon became a well-known centre for the production of fire-arms and great quantities of ordnance, often of the Turkish monstrous type, were currently manufactured in the Moghul Empire.

Despite some local successful achievement, however, no place ever developed manufactures that would even faintly compare with European products. Western guns were always much superior to any non-European product and their superiority was universally recognized. Europe never lost her formidable advantage in armament production – neither quantitatively nor qualitatively. Despite the fact that the 'know-how' was broadcast by renegades, Jesuits, and official missions of 'technical assistance', non-European countries never succeeded in filling the vast technological gap that separated them from Europe. On the contrary, in the course of time the gap grew conspicuously larger.

'It might be supposed,' wrote an eminent scholar,

> that an ignorant man, some edible materials and a cookery book compose together the necessities of a self-moved activity called cooking. But nothing is further from the truth. The cookery book is not an independently generated beginning from which cooking can spring; it is nothing more than an abstract of somebody's knowledge of how to cook: it is the stepchild, not the parent of the activity. The book, in its turn, may help to set a man on to dressing a dinner, but if it were his sole guide he could never, in fact, begin: the book speaks only to those who know already the kind of thing to expect from it and consequently how to interpret it.[68]

Commenting upon this tasty piece of wisdom, another scholar remarked that:

66. Brown, *Impact of Fire-arms*.

67. *Hankuk Haeyang Sa*, pp. 219–25 and Boots, *Korean Weapons*, pp. 22–3.

68. Oakeshott, *Education*, p. 15.

At first sight the problem might appear to be merely one of introducing new methods of production and the instruments, tools or machines appropriate thereto. But what is really involved is a vast change in social beliefs and practices. . . . [Technical knowledge is] the expression of man's response to the changing problems set by the environment and by his fellow men. . . . For meeting any new situation, new thoughts, new aptitudes, new action will be required. But knowledge has to grow: capital has to be created afresh on the basis of continuous experiment, and new hopes and beliefs have to evolve. It is because all these new activities are not independent of the existing institutions into which they have to be fitted, and which have in turn to be adjusted to them, that the process of change is so complex and, if it is to proceed harmoniously, necessarily so slow.[69]

To admit the new role of field artillery in battles of movement and to adopt new strategies accordingly, the Mamluks had to sacrifice the role and prestige of their feudal cavalry, namely the social position and prestige of the dominating class. This in its turn presupposed the disruption of feudal structures and a profound social revolution for which the kingdom was totally unprepared.[70] Before accepting Western techniques, the Chinese had to undergo 'a wholesale change of the world-view, a Copernican revolution of a minor order'.[71] Thus powerful socio-cultural factors opposed the assimilation and diffusion of Western technology. In Europe the situation was vastly different. The European knights of the early Renaissance nourished ideas in regard to fire-arms which were not different from those of the Mamluk horsemen, but by 1500 European affairs were coming more and more under the control of new social groups that had a taste for organization rather than splendour, for efficiency rather than gallantry. And such groups could count on an increasingly numerous class of craftsmen with a taste for mechanics and metallurgy. The very factors that had originally favoured the development of the new technology continued to operate and fostered its further progress powerfully: as has been indicated in the previous chapter, European shipbuilding and manufacture of ordnance moved rapidly ahead during the centuries that followed the first direct contact of the Portuguese with the peoples of Asia.

69. Frankel, *Economic Impact*, pp. 22–4.
70. Ayalon, *Mamluk*, p. 61.
71. Mu, *Hundred Flowers*, p. 17.

It has also to be said that with few exceptions, when an innovation is first introduced, its advantages over established traditions are not always very obvious. The first European field guns were certainly not conspicuous for their efficiency. The attitude of the Turks toward early field artillery, as the attitude of the Venetians toward the early galleons, cannot be simply discarded as a piece of human stupidity. At their first appearance, innovations are less valuable for their actual advantages than for their potential of future developments and this second quality is always very difficult to assess.

The result of the interplay of all these and other factors and circumstances, whatever their respective weight, was one and unequivocal. After the end of the fifteenth century the original 'disequilibrium' between Europe and the rest of the world grew larger instead of levelling out. And for the less 'developed' countries things turned progressively for the worse.

Chapter 3

Epilogue

When Vasco da Gama dropped anchor in the harbour of Calicut, a native asked what the Portuguese were looking for in Asia. The answer of da Gama allegedly was 'Christians and spices'. When Albuquerque attacked Malacca in 1511 he told his officers that they had to exert themselves to the utmost in the coming battle because of two reasons: 'the great service we shall perform to our Lord in casting the Moors out of the country and quenching the fire of the sect of Mahomet . . . and the service we shall render to the King Don Manoel in taking this city because it is the source of all the spiceries and drugs.' Bernal Diaz speculating about the motives that had driven him and his like to the Indies, wrote that they had left Europe 'to serve God and his Majesty, to give light to those who were in the darkness and to grow rich as all men desire to do'.

Through the idea of mission and crusade the conquistadores succeeded where the medieval merchants failed and were able to reconcile the antithesis between business and religion that had plagued the conscience of medieval Europe. One has no reason to doubt the sincerity of their statements, but one may wonder about their realism and the validity of their rationalizations. That the Europeans were more often than not imbued with religious zeal and intolerance is a fact that does not need to be proven. But it is doubtful whether the religious element was as relevant among the motives that drove people overseas as it was among the forces that helped them once they were there. Religious convictions nourished boldness in battle, endurance through ordeals, truculence after victory. But, missionaries apart, when the Europeans undertook the perilous journey, they were dreaming more about Mammon than about lost souls to enlighten. Ogier Ghiselin de Busbecq, the sixteenth-century diplomat, was an inveterate pessimist, but he cannot have been far from the truth when he wrote that for the 'expeditions [to the Indies and the Antipodes] religion supplies the pretext and gold the motive'.

European expansion after 1400 can hardly be depicted as an extension of the Crusades. It cannot be described as the result of Malthusian pressures either. Recurrent epidemics were constantly checking population growth and no population pressure of any relevance was felt in Europe till the second half of the eighteenth century. On the other hand the number of Europeans overseas remained very exiguous until the nineteenth century. Those who left Europe were few. Those who arrived at their destination were half as many. And a large proportion of those who survived the ordeals of the journey and the dangers of life overseas returned to Europe as soon as they could.

European expansion was essentially a commercial venture, and the fact that the colonial policies of the European powers had a very pronounced mercantile tone was the natural consequence of the basic motives behind that expansion. King François I of France was flippantly rude when he styled King Manoel of Portugal '*le roi épicier*' but he was historically correct. One may only add that the States General of Holland and the rulers of England and Spain were no less favourably inclined to 'grocery' than Dom Manoel.

A wide range of economic opportunities magnetized the Europeans overseas. The spice trade, of course, always gave promise of lucrative results. But there was more than that. The Portuguese became increasingly interested in the spice trade toward the end of the fifteenth century. Earlier in the century they crept along the African coast looking for ivory, ebony, slaves, gold, grain and fish. In the sixteenth and seventeenth centuries, when the Europeans had established themselves in the Indian Ocean and the China seas, they did not limit themselves to the spice trade. They were at that time interested in a wide range of commodities from saltpetre to copper, silk and porcelain. The current textbooks of economic history are wrong when they give the impression that the only activity of the European adventurers in Asia was that of supplying the West with Eastern products. The Portuguese, the Dutch and the English were the middlemen in a vast network of commercial activity among Asian nations and a part of European imports were paid for with the income derived from invisible exports of shipping and commercial services.[1] Opportunities were many, risks high but profits higher.

Religion supplied the pretext and gold the motive. The technological progress accomplished by Atlantic Europe during the fourteenth and fifteenth centuries provided the means. As has been indicated in the Prologue, the 'motives' were already active in the Mediterranean Europe as early as the thirteenth century. The Italians and the Catalans were economically very advanced but they did not succeed in out-flanking the Moslem blockade because they were not supported by adequate technology. They made some use of the energy of the wind and later that of gunpowder, but only in a subsidiary way. Essentially they relied on human muscular energy for movement and fighting. But a crew could hardly master the ocean by the use of muscular energy, and when confronted with an enemy it had to yield to superior numbers if the fight were decided by a final mêlée. The link between the Mediterranean and Atlantic developments was Columbus. He had to borrow 'Atlantic vessels, Biscayan sailors and Portuguese nautical techniques'.[2] His role and relevance through the genesis of the project was that of agent of Genoese capital.[3] The contribution of the Mediterranean world to European expansion at the end of the fifteenth century was financial and commercial rather than technological.

The gunned ship developed by Atlantic Europe in the course of the fourteenth and fifteenth centuries was the contrivance

1. The Europeans were very actively engaged in bringing Japanese silver to China (Boxer, *The Great Ship*, *passim*), Japanese copper to China and India (Glamann, *Trade*, pp. 175–6 and *passim*, and Glamann, *Japanese Copper*, *passim*), cloves to India, Indian cotton textiles to South East Asia (Raychaudhuri, *Coromandel*, especially chapt. 8), Persian silk and carpets to India and Japan (Vincenzo Maria, *Viaggio*, pp. 111–12). 'Invisible' exports were vitally important for Europe because of the 'one-way' nature of the trade between Europe and Asia: on this point, see below pp. 149–52.

2. Godinho, *Découvertes*, pp. 50–51.

3. Cf. Caddeo, *Historie*, vol. 2, appendix F, pp. 346–65.

that made possible the European saga. It was essentially a compact device that allowed a relatively small crew to master unparalleled masses of inanimate energy for movement and destruction. The secret of the sudden and rapid European ascendancy was all there: in the skill acquired by Atlantic nations in the use of the sailing ships and in their having understood that 'sea fight in these days come seldome to boarding or to great execution of bows, arrows, small shot and the sword but are chiefly performed by the great artillery.'[4]

When the sailing vessels of Atlantic Europe arrived, hardly anything could resist them. As Albuquerque proudly wrote to his King in 1513, 'at the rumour of our coming the [native] ships all vanished and even the birds ceased to skim over the water.' This was rhetorical but not absurd. Within fifteen years after their first arrival in Indian waters the Portuguese had completely destroyed the naval power of the Arabs and their King proudly styled himself 'Lord of the Conquest, Navigation and Commerce of Ethiopia, Arabia, Persia and India'. Meanwhile business and technology were rapidly advancing in Europe and before the non-Europeans had absorbed the shock of the first contact with the Atlantic vessels, more efficient and more numerous vessels arrived. The caravels and the carracks were followed by the galleons. The Portuguese fleets were followed by the vastly more formidable fleets of the Dutch and the English. The arrival of the new invaders coincided with the outbreak of bloody struggles among the whites. But if the Europeans were tragically divided, their opponents were no more united and they proved unable to take full advantage of the fratricidal quarrels of the Europeans.

The relative advantage of the Europeans was on the seas. On land they remained for a long time highly vulnerable. As has been repeatedly pointed out in the previous chapters, the Europeans were unable to produce an effective, mobile field artillery until the fourth decade of the seventeenth century. Before that date, their land artillery could be moved only with great difficulty. Moreover, its rate of fire could be easily overcome by masses of people. This was a serious drawback, especially

4. See above, chapter 1, footnote 75, p. 71.

overseas where the Europeans were few, and their opponents were many.

Francisco de Almeida pointed out in the sixteenth century that sea power was the key to the situation. 'Let it be known,' he wrote to his King, 'that if you are strong in ships the commerce of the Indies is yours, and if you are not strong in ships little will avail you any fortress on land.' This was also the advice of Albuquerque who wrote to the King that 'if once Portugal should suffer a reverse at sea, your Indian possessions have not power to hold out a day longer than the Kings of the land choose to suffer it.'[5] Cornelis Nieuwenroode, chief of the Dutch factory of Hirado, expressed similar feelings when he wrote in 1623 that the Dutch had 'barely sufficient force to set ashore, unless under the protection of the ship's cannon'.[6] On the other hand, the Europeans were not the only ones to realize how precarious was their position ashore. The Asians knew it too. In 1614 viceroy Chang Ming-kang presented a memorial to the throne in which one reads the following passage:

> Some people are of the opinion that the Portuguese should be removed to Lang-pai or should be permitted to trade with us only aboard their ships which should remain in the open sea. In my opinion we should not resort to the force of arms without duly weighing the consequences. As Macao is within the boundary of our land and forms a part of the district of Hsiang-shan our military forces can watch over the foreigners by just guarding the surrounding sea. We shall know how to put them at death's door as soon as they cherish any disloyal design. Now if we move them to the open sea, by what means could we punish the foreign evil-doers and how could we keep them in submission and defend ourselves against them?[7]

Actual instances of the vulnerability of the European on land are easy to assemble. In the 1620s Abraham van Uffelen the governor of the Dutch factories in Coromandel acted very arrogantly toward the local ruler. When the King decided to strike, van Uffelen could not offer any effective resistance and met with a miserable fate in the prisons of Golconda. In 1638

5. For Almeida cf. Boxer, *Portuguese in the East*, p. 209 and Ballard, *Rulers*, pp. 68*ff*. For Albuquerque cf. his *Commentaries*, vol. 3, pp. 259–60.

6. Boxer, *Fidalgos*, p. 90.

7. Quoted by Chang, *Trade*, p. 120.

Goa would certainly have been lost to the Marathas had not the Great Moghul come to the rescue of the Portuguese. When the Tokugawa Shoguns decided to close Japan to Western influence and decreed the expulsion of the Europeans from the country, the Europeans had no other choice than to comply with the order and leave the country as quickly as possible.

The particular situation by which the Europeans were vulnerable on land while they looked formidable and impregnable on their vessels, explains both the characteristics of European expansion and a curious paradox. The paradox is this: while Europe was boldly expanding overseas and was aggressively imposing her predominance over the coasts of Asia, Africa and the Americas, on her eastern border she was spiritlessly retreating under the pressure of the Turkish forces. Northern Serbia was invaded by the Turks in 1459. Bosnia-Herzegovina in 1463–6. In 1470 the Venetians lost Negroponte. After 1468 it was the turn of Albania. In 1526 the army of King Louis of Hungary was massacred at Mohacs. In 1529 Vienna was besieged. In 1531 Hungary was ravaged and Suleiman retreated only because of the threat from Persia. In 1566 European forces were again retreating in Hungary. In 1596 they were badly beaten at Keresztes. As late as 1606 Western diplomats in Constantinople were fearing further attacks from the Turks and were worrying about the fate of Vienna. In 1683 the Turks were again menacing Vienna.

Thanks to the revolutionary characteristics of their man-of-war, it took only a few decades for the Europeans to establish their absolute predominance over the Oceans. Because their advantage resided in their men-of-war, for almost three centuries their predominance was confined to the seas.

No serious attempt was ever made to make an inroad into Asia and extend territorial conquest. The strategy conceived by the great Albuquerque was strictly followed for more than two centuries after him and apart from a small group of vociferous hot-heads, Europeans generally felt that any attempt to extend their control over Asian hinterlands had no chance of success. In Africa (I mean 'Africa Nigra', south of the Sahara desert) the native populations were by far less numerous and technologically more primitive than in Asia, yet the position of the Europeans ashore was no less precarious. Furthermore, geophysical conditions created an insurmountable barrier to white penetration. The Europeans of the Renaissance had learned to master the energy of the wind and of gunpowder but they were still rather crude in the more general mastery of the adverse forces of nature. 'It seems,' wrote a Portuguese chronicler in the sixteenth century, 'that for our sins or for some inscrutable judgement of God, in all the entrances of this great Ethiopia that we navigate along, He has placed a striking angel with a flaming sword of deadly fevers who prevents us from penetrating into the interior to the springs of this garden, whence proceed these rivers of gold that flow to the sea in so many parts of our conquest.'[8] The few Europeans who ventured into the hinterland were rapidly killed or incapacitated by malaria, tropical fevers, disease and lethal climate. As late as 1876 only ten per cent of the African continent was under white occupation. In the Americas things turned out to be exceptionally more favourable to European invasion. The geophysical conditions of most parts of the continent were not forbidding. The country was not very densely populated and the natives were technologically very primitive. Moreover they turned out to be highly susceptible to European infectious diseases and the spread of deadly epidemics further weakened their already meagre possibilities of resistance. Last but not least, ruthless minorities such as the Aztecs in Mexico were cruelly exploiting large masses of people who were bound to welcome and help any stranger as a liberator.[9] This exceptional combination of favourable circumstances made possible the conquest of vast American territories. The European success and performance however should not be exaggerated. Claims to territorial sovereignty should not be taken as equivalent to actual conquest and until the eighteenth century the areas that came under permanent and effective European control with few exceptions were close to the sea.

Maps show better than any verbal description that until the eighteenth century European possessions the world over consisted mostly of naval bases and coastal strongholds. It is not

8. Quoted by Boxer, *Four Centuries*, p. 27.

9. Simpson, *Encomienda*, pp. vii–viii; and Simpson, *Many Mexicos*, pp. 22–3; Penrose, *Travel*, p. 126.

customary when drawing maps to paint the sea-areas with the colour of the predominant nations. Yet only such a practice would give a correct idea of the nature and extent of European dominance and the role of the Atlantic European nations as world powers during the first centuries of the modern age. Within a few years after the arrival of the first European vessels in the Indian Ocean it became mandatory for non-European vessels to secure sailing permits if they did not want to be blown up by European guns. The oceans belonged to Europe.

The eighteenth century marked the beginning of a new phase. As indicated in Chapter 1, European gun-founders succeeded in producing effective field artillery just before the middle of the seventeenth century. At the beginning the 'innovation' was largely applied at home where the Europeans enthusiastically used it to butcher each other on their numerous battle-fields. Even so, it further enlarged the military technological gap between Europe and the rest of the world and the balance of world power became more unbalanced than ever. As with naval artillery and sailing ships, the Europeans rapidly improved upon their new discovery before the non-Europeans were able to absorb it. The disequilibrium grew, therefore, progressively larger and it was no longer limited to sea power.

Atlantic Europe had initiated maritime expansion. Oriental Europe opened the phase of territorial expansion. First the Turkish menace was brought under control in the course of the eighteenth century. Then Russia took the initiative of the counter-attack which she successfully launched in two directions: to the East against the Hordes of the Kazaks and to the South against the Turks. The iron foundries established by the Dutch in the seventeenth century and developed by Peter the Great in the first half of the eighteenth century and the imports of Western artillery were decisive factors in the advance of the Eastern frontier. As G. F. Hudson puts it:

> The collapse of the power of the nomads with so slight a resistance, after they had again and again turned the course of history with their military powers, is to be attributed not to any degeneracy of the nomads themselves but to the evolution of the art of war beyond their capacity of adaptation. The Tartars in the seventeenth and eighteenth

centuries had lost none of the qualities which had made so terrible the armies of Attila and Baian, of Jenghiz Khan and Tamerlane. But the increasing use in war of artillery and musketry was fatal to a power which depended on cavalry and had not the economic resources for the new equipment.[10]

As to the Turks, this is the testimonial by Baron de Tott who witnessed the events as a military expert:

The Turks were slaughtered in every action by the cannon of the Russians and could only avenge themselves for their disasters by accusing the Russians of cowardly artifice. They overpower us, said they, by the superiority of their fire which, in fact, it is impossible to approach; but let them leave their abominable batteries and encounter us like brave men, hand to hand, and we shall soon see whether these infidels can resist the slaughtering sabre of the True-Believers. . . . The Grand Seigneur having been informed that the howitzers had very much annoyed his cavalry, enquired of me what kind of artillery they were, for they were unknown at Constantinople. . . . The destruction or at least the entire dispersion of the Turkish army by the action at Craool, had already induced His Highness to imagine that the quick fire of the Russian artillery was the principal cause of the discouragement of his troops.[11]

Overseas the impact of the new European technological developments was not felt until the end of the eighteenth century because of the difficulty of supplying a large army over long distances from the home base. As late as 1689 the forces of the East India Company were completely routed on land in India. In 1700 the Directors of the Company regarded the idea of conquering territories or planting colonies in the country as 'altogether impracticable in respect of a long voyage, the diseases to which our people are liable in those hot regions and the power, force, and policy of most Indian nations'.[12] The conquest of the country in the second half of the eighteenth century was possible only because of the state of anarchy into which India was plunged after the death of Aurangzeb in 1707 and the defeat of the Marathas at the hands of the Afghans.

The Europeans' effective conquest or control of vast hinter-

10. Hudson, *Europe*, p. 268.

11. De Tott, *Memoirs*, vol. 2, part 3, pp. 10 and 79.

12. Quoted by Thomas, *Mercantilism*, p. 10. On the campaigns of 1685 and 1689 cf. Wilson, *Early Annals*, vol. 1, pp. 102*ff*.

lands came later as one of the by-products of the Industrial Revolution.

European maritime expansion was one of the circumstances that paved the way for the Industrial Revolution. To deny it on the basis that there were no West Indies merchants or East-India adventurers among the 'entrepreneurs' who built factories in Europe is as sensible as to deny any relation between the Scientific Revolution and the Industrial Revolution on the basis that neither Galileo nor Newton set up a textile mill in Manchester. Inter-relationships in human history do not always work so openly and crudely.

On the other hand, there is no doubt that the Industrial Revolution in its turn added great impetus to European expansion. It multiplied the number of the Europeans in absolute terms as well as in relation to the size of non-European population. It provided the Europeans with more powerful weapons and with efficient techniques to master the adverse forces of nature. And it gave Industrial Europe the opportunity to subjugate non-industrial economies through the policy of 'free trade' and the subtle mechanism of 'dual economies'. As Adam Smith saw it, 'in ancient times the opulent and civilized found it difficult to defend themselves against the poor and barbarous nations; in modern times the poor and barbarous find it difficult to defend themselves against the opulent and civilized'.

In the passage by Adam Smith just quoted, readers may perceive traces of a disturbing confusion between 'civilized' and 'technologically advanced', a confusion that, at least in the form in which we are familiar with it, is one of the by-products of the Industrial Revolution. Queen Victoria's gunboats defeated Lin's noble efforts to put to an end the opium trade but this does not imply that the admirals of Queen Victoria were more 'civilized' than Commissioner Lin. If the historical analysis of this book is correct, the technologically more advanced people are bound to prevail regardless of their degree of 'civilization' which is something more difficult to define and assess.

The 'Vasco da Gama era' has now come to an abrupt end. In its uprising against Western predominance, the 'underdeveloped' world seems to emphasize exclusively the importance of the acquisition of Western technology. Since Western predominance has rested on superior technology, the attitude is understandable, but it has tragic implications.

In order to acquire Western techniques, the non-European people had or have to undergo a more profound and general process of 'westernization'. Paradoxically enough, in order to fight against the West they have to absorb Western ways of thinking and doing. As M. Chiang wrote:

> Since we were knocked out by cannon balls, naturally we became interested in them, thinking that by learning to make them we could strike back. We could forget for the time being in whose name they had come, since for us common mortals to save our lives was more important than to save our souls. But history seems to move through very curious ways. From studying cannon balls we came to mechanical inventions, which in turn led us to political reforms; from political reforms we began to see political theories, which led us again to the philosophies of the West. On the other hand, through mechanical invention we saw science, from which we came to understand scientific method and the scientific mind. Step by step we were led farther and farther away from the cannon ball – yet we came nearer and nearer to it.[13]

In this process, the goal is the technique while philosophy and social and human relations are degraded to the role of means. The machine which should serve man, becomes his master. The 'Vasco da Gama era' ends in a nightmare in which men – Westerners and non-Westerners alike – are bewildered by this confusion and the old fancy of the *apprenti sorcier* becomes tragically actual.

13. Chiang, *Tides*, p. 4.

Clocks and Culture

Chapter 1

The European Masters

From the earliest antiquity, man has created various devices to solve the problem of the measurement of time. Sundials were the first solution and they continued to be widely used well into the sixteenth and seventeenth centuries, long after the appearance of mechanical clocks. Their low cost and their precision account for the great variety of types that were developed in the course of the centuries.[1] But the sun does not always shine, and man had to invent other instruments: water clocks and fire clocks[2] were developed in the ancient world; more recently, sand glasses came into use, and in our atomic age they are still used for timing the boiling of eggs.

Of these different devices, the water clock or *clepsydra* seems to have been the one that offered to ingenious craftsmen the greatest opportunity for developing extravagant variations on the

1. In the Berlin Museum is preserved an Egyptian shadow-clock that dates from about 1500 B.C. but similar devices for telling the time must have existed long before that date. In the course of time, many types of sundials appeared: equatorial sundials, horizontal sundials, vertical sundials, declinate sundials, azimuth sundials, portable sundials, etc.

2. King Alfred's biographer, Asser, whose accounts are fairly well authenticated, relates that Alfred made use of special candles as time-keepers (Ward, *Time Measurement*, p. 11). In A.D. 758 Pope Paul I sent King Pepin a present of a few books '*necnon et horologium nocturnum*' (*Codex Carolinus*, p. 513). At the times of King Lothar (A.D. 954–986), a deacon in Verona made an '*horologium nocturnum*' such as 'nobody had ever seen before' (Muratori, *Dissertazioni*, vol. 1, pp. 364–5. Cf. also Belgrano, *Antichi Orologi*, p. 30). King Louis IX (A.D. 1226–70) used candles of given length to measure the time while he was reading (*Vie de Saint Louis*, p. 79) and a similar device was still used by the French King Charles V (A.D. 1364–80) (De Pisan, *Sage Roy Charles*, vol. 1, p. 609). On Eastern fire clocks cf. Bedini, *The Scent of time*.

basic principle. The primitive form of water clock was a stone vessel from which the water was allowed to escape slowly through a small hole, time being indicated by the level of the water within. In the course of time more elaborate *clepsydrae* were made, containing mechanisms that moved automata and struck the hours. It does not take an extreme flight of fancy to imagine that once *clepsydrae* with trains of wheels had been constructed, some craftsman must have thought of the desirability of substituting for the flow of water some other motive power. At the same time there were craftsmen who were bothering their heads with bells and all possible mechanisms for ringing them effectively.

Bells played a prominent role in the life of medieval towns: they ruled the life of the community and their sound lifted 'all things unto a sphere of order and serenity'. Everybody knew their meanings, and bells rang out their messages at all times, telling the hours, announcing fire or an approaching enemy, calling the people to arms or to peaceful assemblies, telling them when to go to bed and when to get out of it, when to go to work, when to pray and when to fight, marking the opening and closing of fairs, celebrating the elections of popes, the coronations of kings, and victories in war. According to a widespread belief, the sound of bells also helped to keep away storms and epidemics. It was a matter of pride for a town, a church, a monastery to have a beautiful bell or peal of bells, and in the course of time, mechanisms were developed to ring bells with greater efficiency: it is not improbable that these contrivances, which were made of toothed wheels and oscillating levers, helped to prepare for the development of mechanical clocks. Finally, there were astronomers, astrologers and others interested in making globes and spheres and in supplying them with movements that would imitate the movements of the stars and the planets. The historian is inclined to fancy that techniques developed in the making of these devices must also have brought craftsmen closer to the making of mechanical clocks.

If a broad historical point of view invites one to stress the continuity and gradualness inherent in the process of technological change, a strictly technological point of view forces onc to emphasize the fundamental difference between water clocks, bell-ringing mechanisms and the like on the one hand, and the mechanical clock on the other. Historians of technology and science have good cause to dramatize the sharp break in the history of horology brought about by the invention of the verge escapement with foliot. In the water clock, the time-keeping is governed by the rate of the flow of water through a hole, and the regulating device is simply the hole through which the water flows. In the mechanical clock, the time-keeping is governed by an oscillator or escapement which controls the unidirectional movement of the motive power and transforms it into a slow, steady and regular motion whose meaning appears on the dial of the clock. The verge escapement with foliot is a most ingenious device and, as has been said, whoever invented it must have been a mechanical genius. The interested reader will find a technical description of the device below in the Appendix; here it suffices to say that the mechanical clock was born when the verge escapement with foliot was invented. Historians have long debated the dating of this invention but the general consensus now places it in the second half of the thirteenth century.[3]

Towns were then booming and the new urban civilization was asserting itself with unprecedented vigour, as yet unhampered by the rigidities that develop in the course of time around institutions, traditions and vested interests. The thirteenth century saw the diffusion of the Universities and of the Gothic cathedrals, the aesthetic revolution brought about by Giotto and Cimabue, the voyages of the Polos to China and the first effort by Europeans to sail along the west coast of Africa, searching for a sea passage to the East; the second half of the century saw the making of the first cannon.[4] It was not entirely by chance that the mechanical clock and the cannon appeared at approximately the same time. Both were the product of a remarkable growth in the number and quality of metal workers, and as we shall see

3. Cf. Thorndike, *Invention of the mechanical clock*, pp. 242–3; Lloyd, *Outstanding clocks*, pp. 1–8; Needham, and ass., *Heavenly Clockwork*, pp. 195–6; Zinner, *Frühzeit der Räderuhr*, pp. 8–11.

4. The first mention of cannon in Europe occurs in a Florentine document of 1326. From that document we learn that the town council had decided to buy some bronze guns and iron gunshot for the defence of the city. If cannon were already in use and could be bought in the 1320s, one may guess that their 'invention' occurred some time toward the end of the thirteenth century.

later, many of the early clockmakers were also gun-founders. The simultaneous appearance of the gun and the mechanical clock was both a testimony to the character of European development and a forecast of things to come.

It has recently been suggested that the mechanical clock had a Chinese origin, and those who favour this theory point to a remarkable piece of machinery built in China toward the end of the eleventh century. This contraption consisted essentially of a huge water mill that was equipped with some kind of escapement. The Chinese escapement however had nothing in common with the European verge-and-foliot device. The suggestion therefore is that stories of the 'heavenly clockwork' must have reached Europe and 'perhaps ... some scholar-craftsman puzzling over the tale that in the East they used a set of oscillating levers, tripping and holding back a wheel so to regulate its turning ... made his own oscillatory device'.[5] Allowing their fancy an equally free rein, some Arab circles maintain that Dante's *Divine Comedy* was an imitation of the celestial fantasy written three centuries earlier by Abu'l-Ala al-Ma'arri.

The development of urban life in the West vastly increased the need for measuring time. No less important, however, was the fact that the Europeans were increasingly inclined to satisfy such needs by the use of mechanical contrivances. In the course of the fourteenth century mechanical clocks became progressively more numerous in Europe, and very soon they were equipped with mechanisms for striking the hours. A clock made of iron was installed at the church of St Eustorgio in Milan in 1309. The cathedral at Beauvais had possibly a clock with a bell before 1324. In 1335, according to an Italian chronicler,[6] the church of St Gothard in Milan had

> a wonderful clock, with a very large clapper which strikes a bell twenty-four times according to the twenty-four hours of the day and night and thus at the first hour of the night gives one sound, at the second two strokes ... and so distinguishes one hour from another which is of greatest use to men of every degree (*quod est summe necessarium pro omni statu hominum*).

5. Needham and ass., *Heavenly Clockwork*, p. 197.

6. Flamma, *Opusculum* in Muratori, *Rerum Italicarum Scriptores*, vol. 12, col. 1011; Belgrano, *Antichi Orologi*, pp. 32–3.

The monastery of Cluny had a clock in 1340 and by 1359 the Cathedral of Chartres had two. In Padua a public clock that 'through days and nights tells the twenty-four hours automatically (*sponte sua*)' was installed in 1344. Public clocks that rang the hours appeared in Genoa for the first time in 1353, in Bologna in 1356 and in Ferrara in 1362. In 1370 Charles V of France installed a clock that struck the hours on one of the towers of the royal palace, and so pleased was he with this that he had similar clocks installed at the Château de Vincennes and the Hôtel Saint-Paul. Being afraid that their sound could not be heard by everyone in town, he also ordered all the churches in Paris to ring their bells '*par pointz à maniere d'orologe*' when the royal clocks struck the hours. Thus everyone could know the time '*luise le soleil ou non*'.

A clock, especially a big public clock, was very expensive in those days. It was expensive to build, and its maintenance,[7] which generally included the regular wage of a specially appointed 'governor'[8] was a severe drain on local finances. The decision to install or not to install a clock was often the result of long and heated debates, but it appears that in general people took great

7. Froissart wrote in his poem *Li Orloge amoureuses*, pp. 79–80,

li orloge ne poet aler de soi
se il n'a qui le garde et qui en songne
pour ce il faut à sa propre besongne
un orlogier qui tart et tempre
diligamment l'administre, et attempre
le plons relieve et met à leur devoir.

In some cases the craftsman who built the clock was appointed governor of the same; in other words when buying a clock one had to buy its maker too.

8. As Gélis, *Horlogerie ancienne*, p. 48, writes, 'the office of a governor of the clock was not a sinecure. Often the governor had to wind up the clock twice a day and he had therefore to climb twice a day to the top of the clock tower; he had very frequently to grease the machine, because the gears were not so smoothly and precisely constructed; he had finally to reset the hand (or the hands) of the clock almost every time this was being wound, because the clock lost or gained much time in the course of half a day'. Master John de Lendenaria knew pretty well that it was not an easy task to be governor of a clock. In the last days of January, 1436, the poor fellow fell from the top of the staircase in the clock tower in Castronovo near Ferrara. Weeks later he was still in bed, with many of his bones broken, his body aching all over: '*adhuc resupinus cum doloribus maximis jacet in lecto*' (Campori, *Orologieri*, p. 262).

pride in their public clocks and considered them essentially useful things. The Milanese chronicler was convinced that the clock in the church of St Gothard 'was of greatest use to men of every degree'. Having in mind the clock installed by Charles V on the Royal Palace in Paris, Froissart wrote:

L'orloge est, au vrai considerer
un instrument très bel et très notable
et s'est aussi plaisant et pourfitable
car nuit et jour les heures nous aprent,
par la soubtilleté qu'elle comprent
en l'absense meïsme dou soleil
dont on doit mieuls prisier son appareil.[9]

In 1473, Bartolomes Manfredi asserted that the complicated cosmological indications of the public clock in Mantua (Italy), served the purpose of showing 'the proper time for phlebotomy, for surgery, for making dresses, for tilling the soil, for undertaking journeys and for other things very useful in this world'.[10] In 1481 a petition presented to the Town Council of Lyons stressed the fact that in the town 'is sorely felt the need for a great clock whose strokes could be heard by all citizens in all parts of the town. If such a clock were to be made, more merchants would come to the fairs, the citizens would be very consoled, cheerful and happy and would live a more orderly life, and the town would gain in decoration.'[11] In an age in which nobody, or very few, had a portable clock, the usefulness of the public clock striking the hours was unquestionable; but practicality was not always the only motivation. Some towns rivalled others for the distinction of having, as a fifteenth-century French document put it, '*relotgium magnum sufficiens et honorabile ad honorem villae*'.[12] About 1380, the Town Council of Lyons decided to install on one of the bridges a tower with a clock similar to one on a bridge in Paris: '*prout et quemadmodum edificate sunt Parisiis turris et horologium desuper existens*'.[13] In the 1420s the Town Council of Romans (France) decided to build a very beautiful clock 'without any

9. Froissart, *Orloge amoureuses*, p. 53.
10. Davari, *Notizie storiche*, p. 220.
11. Vial and Côte, *Horlogers Lyonnais*, p. 4.
12. Reverchon, *Histoire de l'horlogerie*, p. 62.
13. Vial and Côte, *Horlogers Lyonnais*, p. 2, *n*. 5.

regard to expenses' ('*sans regarder à la dépense*'). In 1557 the inhabitants of Montélimar (France) decided to have a clock similar to that of Romans: '*a la forma d'icelluy de Romans*'.[14]

Thus a combination of civic pride, utilitarianism, and mechanical interest fostered the diffusion of the clock despite its relatively high cost. Since more clocks were being made, the proficiency of the makers improved, and by the end of the fourteenth century, clocks which struck the hours and the quarters had been built,[15] but this fact should not mislead us regarding the precision of these early time-keepers. As has been said, early clocks 'embodied a strange combination of brilliance in conception with a deficient technique of construction'. Throughout the fourteenth and fifteenth centuries, most clocks (if and when they worked), lost or gained much time in a day. On the other hand, contemporary requirements for precision were low: thus it was generally thought unnecessary to provide clocks with a minute hand. In 1389 the city of Rouen granted a stipend to the wife of a clockmaker for turning the clock two or three times a week; in all likelihood, she had to reset the hand to correct the time. In 1387, King John of Aragon decided to employ two men to strike the bells of the clock in his castle of Perpignan since the clock itself failed to strike correctly. The people of Paris referred to the clock of the royal palace in the following rhyme: '*l'horloge du palais, elle vas comme il lui plait*'. There is no doubt that until at least the sixteenth century even the best clocks kept the time only roughly and had to be regularly reset by sundials. As late as 1641 the town council in Dijon, noticing in desperation that no one of the public clocks agreed with the others, gave the peremptory order that they should be put in accordance '*suivant le cours du soleil*'.

The most striking occurrence in the early history of clocks is that while medieval craftsmen did not improve noticeably in precision, they soon succeeded in constructing clocks with curious and very complicated movements. It was easier to add wheels to wheels than to find better ways to regulate the escape-

14. Fillet, *Horloges publiques*, pp. 104–5.

15. This was the case of the clock of the Cathedral at Wells in England. Cf. Britten, *Old clocks*, p. 12.

ment. On the other hand complicated movements had quite a popular appeal and most people believed that a correct knowledge of the conjunction of the heavenly bodies was essential for the success of human enterprises. One of the most remarkable pieces in this regard was the clock made about 1350 for the cathedral of Strasbourg. Enormous in size, it included a moving calendar and an astrolabe whose pointers indicated the movements of the sun, moon and planets. The upper compartment was adorned with a statue of the Virgin before whom at noon the Three Magi bowed while a carillon played a tune. On top of the whole thing stood a cock which, at the end of the procession of the Magi, opened its beak, crowed in a sinister way and flapped its wings. In Bologna around the middle of the fifteenth century Master Giovanni Evangelista da Piacenza and Master Bartolomeo di Gnudolo built an impressive clock on the *Palazzo del Comune*, with a trumpeting angel, a large number of Saints and other holy dignitaries and a procession of the Magi who payed their respects to the Virgin and Child. The most remarkable thing, however, was the astronomical section of the clock. Built under the supervision of that great humanist, Cardinal Bessarion, the astronomical section allegedly showed 'a central globe of fire around which the sun, the moon, the earth, the planets and the skies rotated harmoniously'.[16] This arrangement was based on the cosmological views held by the disciples of Pythagoras and it was in open contrast with the Ptolemaic theories prevailing in the fifteenth century. The famous clocks of Orvieto and Reggio in Italy, of Wells in England, of Lund in Sweden, of Lübeck in Germany, and of Berne in Switzerland were not quite as impressive as the clock of Strasbourg or so unorthodox as the clock of Bologna but they were, none the less, extraordinary pieces. For the sake of beauty and civic pride, complicated movements were sometimes added to existing clocks. This was done in Parma (Italy) in 1431 and the chronicler, obviously class-conscious, wrote that the new contrivances told the hours

16. Rubbiani, *L'orologio del comune di Bologna*, pp. 349–66. The fact was truly exceptional, especially if one considers that even in the two centuries after the Copernican discoveries it was still customary to use the 'Ptolemaic' for astronomical dials as it was easier to use it to indicate the hour as well as other things. As a 'reference', which is the first requisite of a dial, the 'Copernican' system is useless.

'to the commoners' ('*al popolo*') and the position of the moon to 'those who can understand' ('*agli intelligenti*').[17] In 1510 an elaborate jack-work was fitted in the clock on the belfry of Ghent (Flanders): a man (Adam) struck the hours, a woman (Eve) struck the half-hours and a snake moved around obviously trying to induce Adam and Eve to indulge in other and more synchronized kinds of movements. The masterpiece of the Middle Ages, however, was the astronomical clock made about 1350 by Giovanni de'Dondi, probably with the help of his father Jacopo. It was only incidentally a timepiece: as a matter of fact it was more remarkable for its astronomical work than for its actual clockwork. It included the celestial wanderings of sun, moon, and five planets and provided a complete and perpetual calendar. Philippe de Maizières, who saw and admired it, described it as follows:

There is in Italy today a man exceptionally versed in philosophy, medicine and astronomy who is by general consensus recognized as the greatest living authority in these three sciences. His name is master John de Dons and he lives in the city of Padua. Because of his great knowledge in astronomical matters his family name is generally forgotten and he is called master John of the clocks. He is now with the Count of Virtú[18] and he gets a salary of two thousand gold florins a year for his knowledge in the three fields. This master John of the clocks has produced famous works in the three sciences, works that are held in great repute by the great scholars in Italy, Germany and Hungary. Among other things, he has made an instrument, called by some a sphere or clock for the celestial movements; which instrument shows all the movements of the signs and the planets with their circles and epicycles, and differences, and each planet is shown separately with its own movement in such a way that at any moment of the day or of the night one can see in which sign and to which degree the planets and the great stars appear in the sky. The sphere is constructed in such a subtle way that in spite of the fact that there are so many wheels that they cannot be counted without taking the clock to pieces, all goes with one weight. So great is the marvel that great astronomers come from distant places to admire his work.... In order to have his sphere done as he had it in his subtle mind, the said master John

17. Belgrano, *Antichi Orologi*, p. 40.

18. The reference is to Giangaleazzo Visconti, Duke of Milan, Prince of Pavia and Count of Virtu.

actually forged it with his own hands out of brass and copper, without the help of anyone, and he did nothing else for sixteen years.[19]

The masterpiece was eventually placed in the library of the Visconti castle at Pavia, but after the death of John of the clocks, nobody knew how to take care of it. The clock was destroyed in the course of time, but we know its mechanism thanks to an accurate description made by Master John himself and handed down to posterity. Nowadays the admiration for the horological masterpiece of Master John has not yet subsided. Alan H. Lloyd, an expert in such matters, thinks that 'had a similar clock been devised and made today in spite of all the knowledge and machinery that would be available to the maker, he would be regarded as an outstanding man',[20] and Professor L. White adds:

> [Giovanni's] sense of the interrelation of moving parts showed genius: to provide for the elliptical orbits of the Moon and Mercury (as required by the Ptolemaic system) he produced elliptical gears and likewise made provision for the observed irregularities in the orbit of Venus. In complexity and refinement Giovanni's gearing goes enormously beyond anything which survives from earlier technology, including the fragments of the Hellenistic planetarium found in the Aegean Sea. In this aspect of machine design, the fourteenth century marks an epoch.[21]

Most of the clocks mentioned in the previous pages were public clocks. Since mechanical time-keepers were very expensive, it is only natural that their early diffusion should be greater in the public sector where the necessary funds could be raised through voluntary or compulsory joint financial effort. Domestic clocks, however, were not completely lacking. When that extraordinary collector of *objets d'art* Charles V of France died in 1380, the officials who made the inventory of the 3,985 items of his collection found among them 'one clock all made of silver and with no iron, that had belonged to the late King Philip the Fair, with two weights covered with silver and filled with lead'.[22]

19. From *Le Songe du Vieil Pélerin adressant au Blanc Faucon à bec et pieds dorés*, by Philippe de Maizières, counsellor to the King Charles V of France. The passage was reproduced in *Histoire de l'Académie Royale des Inscriptions*, 16 (1751), pp. 227–8.

20. Lloyd, *Outstanding clocks*, p. 24.

21. White, *Medieval Technology*, p. 126.

22. Labarte, *Inventaire*, pp. 278–9, item no. 2598.

This must have been a domestic clock driven by weights and it must have been constructed before 1314, the year of Philip the Fair's death. The 'sphere' constructed by Dondi, no matter how exceptional for its astronomical intricacies, was also essentially a domestic clock. Other examples of domestic and portable clocks can be found if we search the dust of the archives, but it can be said with a good deal of plausibility that they were extremely rare until the middle of the fifteenth century. For one thing, they cost too much; for another, a clock was considered to be a very intricate, delicate, and unpredictable machine which required the continual care of a 'governor' who had to 'rule, set, guide and keep it'. Even Charles V, who had more than one clock in his collection of *objets d'art* and who had installed public clocks in some of his palaces, relied on specially graduated candles to measure the time in his own room. Domestic clocks, said Christine de Pisan to whom we are grateful for the information, '*encore n'estoyent communs*'.[23]

Yet the diffusion of public *horologia*, which had made people accustomed to clocks, paved the way for the diffusion of domestic clocks. After the first craze for the big and elaborate public *horologia* had subsided, one has the impression that more time and ingenuity were devoted to the making of smaller, portable pieces. The forces of the market were slowly but irresistibly indicating that it was in this direction that in the course of time, production could find larger outlets.

As long as weights were the only motive power, house clocks were not easily movable; they had to stand on brackets or be spiked to the wall. In order to construct easily portable clocks, a new kind of motive power had to be devised. According to a contemporary of Filippo Brunelleschi, the great Italian architect was very interested in clockwork and by 1410 was constructing clocks driven by 'various and diverse kinds of springs'.[24] The word that the Italian biographer used, '*molle*', undoubtedly means 'springs' but one cannot say that his description sounds very technical. On the basis of this passage alone, one would hesitate to conclude that by 1410 spring-driven clocks had

23. De Pisan, *Charles V*, vol. 1, p. 609.

24. Morpurgo, *Ruote o molle*, pp. 31–2. Cf. also Morpurgo, *Dizionario*, p. 33.

appeared in addition to weight-driven ones. However, a French illuminated manuscript dating from about 1440–50 portrays in great detail the mechanism of a portable clock that was spring-driven.[25] The same miniature and other and earlier documents illustrate an ingenious device, the *fusée*, which had the function of preserving a more or less constant torque throughout the running time of the spring. Available evidence seems therefore to indicate that the use of the spring in timepieces dates back at least to the first decades of the fifteenth century. The importance of the invention can hardly be overestimated, for it made possible the construction of easily portable clocks and, later, the construction of the watch.

Throughout the second half of the fifteenth century, domestic weight-driven clocks were still very rare and spring-driven clocks even more so. About 1450 an important personage of the Burgundian court who owned a spring-driven clock had it painted in the background of his portrait.[26] In 1481 King Louis XI of France paid £16 10s. (*tournois* money) 'to Jehan of Paris, clockmaker, for a clock that has a dial and strikes the hours, complete in all its parts; which clock the King bought to carry with him in all places where he shall go'.[27] That the king was very proud of this piece is clearly indicated by the fact that he too had it painted in one of his portraits.[28]

It was in the sixteenth century that domestic clocks and watches became much less of a rarity.[29] If one compares the European scene with other parts of the globe during the sixteenth and seventeenth centuries, one is impressed by the fact that there were many more craftsmen and merchants in relation to total population in Europe than elsewhere. But no less striking and no less important was the fact that in Europe between the few who were very rich and the many who were very poor there existed a relatively large group of urban dwellers – merchants, lawyers, notaries, doctors, apothecaries and the like – who could afford

25. Michel, *Horloge de Sapience*, pp. 297–8.
26. Lloyd, *Outstanding clocks*, pp. 30–32.
27. Douët-D'Arcq, *Comptes de l'hotel des rois*, p. 388.
28. On this portrait cf. Gélis, *Horlogerie ancienne*, pp. 4–6.
29. Pocket watches appeared toward the end of the fifteenth or at the very beginning of the sixteenth century: on the question cf. Morpurgo, *L'orologio tascabile;* Reverchon, *Histoire de l'horlogerie*, pp. 67–8.

decent houses, good clothes, and a few amenities of life. The two facts, one acting on the side of supply and the other on the side of demand, were firmly interlocked: each one existed only because of the other. In sixteenth- and seventeenth-century Europe there were not only craftsmen who could make timepieces but also a relatively large number of people who could buy them. Thus the production of clocks and watches progressively expanded.

What had happened previously to the public weight-driven clocks happened in a more or less similar way to spring-driven portable clocks and watches: after their first appearance, craftsmen spent most of their time and effort in constructing pieces with complicated calendrical and astronomical movements or in devising curious shapes. As far as precision is concerned, clocks and watches remained relatively poor time-keepers throughout the sixteenth century and the first half of the seventeenth. Of course, some progress was made: in the course of the sixteenth century the minute hand appeared more frequently, and one can find in museums some outstanding clocks of that period that keep time with a remarkable degree of accuracy. One may recall, for instance, the clocks made by Jost Burgi and by Jost Bodeker. However, one should not generalize from a few exceptional specimens and assume these museum-pieces to be a representative sample of the norm. The invention of new and better escapements and the solution of some difficult problems of applied mechanics were necessary in order to make possible the large scale production of ordinary time-keepers possessing a satisfactory degree of precision.

In the fourteenth and fifteenth centuries the demand for clocks was not great enough to allow for the growth of a specialized group of craftsmen. The earliest clockmakers were mostly blacksmiths or locksmiths or gun-founders; in other words, they were craftsmen who knew how to work metals and who on occasion also managed to produce or to repair mechanical time-keepers. At the beginning of the fifteenth century, Jacques Yolens of Lille was 'a clockmaker and a gunner'[30] and master Pierre Cudrifin of Fribourg was '*magister bombardarum et horologiorum*'.[31]

30. Henrard, *Documents*, p. 169.
31. Fillet, *Horloges Publiques*, p. 104.

In the Genoese colony at Caffa (Crimea) in 1455 Ubaldinus de Florentia was '*bombarderius et magister orologii Comunis*'.[32] Henricho, the parson who was appointed governor of the clock at the Church of St Gothard in Milan in 1474 was a reputed expert in matters of bombards.[33] After the end of the fifteenth century Noël Cusin, governor of the clock of Autun's cathedral, made clocks, organs and cannon[34] and in Lisbon Master John, probably from Germany, was a gunner and a clockmaker.[35] Other examples can easily be found in great numbers. This colourful type of versatile craftsman actually survived for a long time in those places where the manufacture of timepieces did not develop into an important economic activity. In Flensborg (Denmark) in the 1550s Gert Merfelden made cannon and clocks and in the first part of the seventeenth century in Randers (Denmark) a craftsman made clocks and shafts for carriages.[36] In Dundee (Scotland) in the 1580s, a Patrick Ramsay who 'did thankful service by his good attention on the knok and steeple' of St Mary's Church was 'smith and gunmaker'.[37] In seventeenth-century Basel (Switzerland) 'locksmith and clockmaker' was a common professional qualification and as late as the 1730s most of the craftsmen who repaired clocks and watches in Berlin were 'locksmiths'.

In the areas of medieval Europe where metal workers were scarce and the mechanically-inclined ones were rare, clocks were often constructed by foreign craftsmen. In Catalonia most of the clockmakers were Jews. In fourteenth-century England the clocks of the cathedrals of Salisbury and Wells were in all likelihood made by foreign craftsmen brought to England by Bishop Erghum, himself a foreigner. In 1368 Edward III granted protection and safe conduct for one year to John Vueman, William Vueman and John Lietuyt of Delft (Holland), 'orlogiers coming into the Realm to practice their art'.[38] When foreigners were not at hand, friars occasionally devoted themselves to the task. In

32. Belgrano, *Antichi Orologi*, p. 54*n*.
33. Motta, *Musici*, p. 529.
34. Babel, *Histoire corporative de l'horlogerie*, p. 46.
35. Pieris and Fitzler, *Ceylon*, vol. 1, p. 299.
36. Liisberg, *Urmagare og Ure*, p. 137.
37. Smith, *Scottish Clockmakers*, p. 125.
38. *Calendar of the Patent Rolls*, Edward III, 4 May 1368.

England Father Richard of Wallingford is credited with the construction of horological devices in the early part of the fourteenth century. In the 1360s a clock was made for the Pope in Avignon by a friar from Venice. In fifteenth-century Sweden, some friars of the Vadstena monastery were interested in horology and the astronomical clock in the Dome at Uppsala was made in 1507 by one of them. In 1537 the town clock at Aberdeen (Scotland) 'was reformed and mended by friar Alexander Lyndsay'.

Germany (using the geopolitical term in its wide medieval connotation)[39] appears to have had a relative abundance of capable smith-clockmakers. The German craftsmen very early acquired a good reputation and were often invited to countries such as France and Italy where native craftsmen were not lacking. In 1370, when Charles V of France wanted to install a public clock in his palace, he brought to Paris a German craftsman, Henry de Vic; in 1407 it was another German, Jehan d'Alemaigne, 'locksmith' who made in Paris 'a small clock (*petite orloge*) to be put in the room of the Queen'.[40] In the fifteenth century, some German clockmakers were working in Milan, Rome and other Italian towns.[41] However, although skilled smith-clockmakers were more numerous in some areas than in others, there were no real centres of clockmaking in Europe until the end of the fifteenth century. Clocks were usually produced where they were consumed, and, if necessary, it was the craftsmen who travelled, not the merchandise. Clockmakers were in demand because there

39. Used in this way, the term 'Germany' includes most of Switzerland, the northern and southern Low Countries, Alsace, Lorraine, Austria and the Tyrol. Cf. also below note 40.

40. On Henry de Vic and his clock at Paris, cf. Vidier, *Horloge du Palais*, pp. 95–6. On Jehan d'Alemaigne and his work for the Queen of France cf. De Laborde, *Notice des émaux*, vol. 2, p. 415.

On German locksmiths-clockmakers working in Besançon in the fifteenth and sixteenth centuries, cf. Sandoz, *Maîtres horlogeurs*, pp. 5 and 7. As late as 1650 German clockmakers were invited to Lyons to help in the construction of an elaborate astronomical clock; cf. Vial and Côte, *Horlogers Lyonnais*, p. 56.

41. On German clockmakers working in Lombardy cf. N.N., *Tedeschi in Milano*, p. 997, *n.* 2; P.P., *L'Orologio dell'Ospedale*, pp. 687–8; Canetta, *Vicende edilizie*, p. 351; etc. On German clockmakers working in Rome and other Italian towns cf. Zinner, *Wurde die Räderuhr in Deutschland oder in Italien erfunden*, pp. 17–22, and Morpurgo, *Dizionario*, *passim*.

were few of them, but they could hardly settle in numbers in any one place because the local demand for their work was easily saturated. For all these reasons, as well as for the fact that they were hardly distinguishable from other metal workers, the clockmakers had no guild of their own. Thus to the very end of the Middle Ages the organization of the craft retained certain primitive features long since abandoned by the textile industry and other trades.

In the course of the sixteenth and seventeenth centuries the greater demand for domestic clocks and watches permitted the formation of settled groups of artisans. First, Augsburg and Nuremberg emerged as centres of clockmaking. Later on, other centres developed such as Blois, Paris, Lyons, Geneva and London. In Paris there were about twenty master clockmakers beside an unknown number of workers and apprentices at the middle of the sixteenth century[42] and about seventy masters in the 1640s.[43] In Blois there were about seventeen masters in 1600 and about forty-five in 1639.[44] In Lyons there were more than ten masters in 1570, about sixteen in 1610, and between forty and sixty just after the middle of the seventeenth century.[45] In Geneva there were more than twenty clockmakers toward the end of the

42. In July 1544 a royal charter was obtained by seven 'master-clockmakers residing in Paris' for incorporating a guild of the clockmakers of Paris (Lespinasse, *Métiers et corporations*, vol. 3, p. 546 and 549–52). In accordance with the royal charter, a list was prepared with the names of the master clockmakers 'who exercise their profession and have their shops in the town of Paris'. The list indicates also the 'masterpieces' that the masters had produced or had to produce in order to be admitted to the guild. The document (*Archives Nationales* Paris, Y 6 [s], fol. 109), dated Sept. 1545, mentions twenty masters. The names of two are written over with the note 'dead without having produced the masterpiece'. Two other names were, however, added later on to the list. In 1549 there were ten clockmakers in the procession that celebrated the coming to Paris of King Henry II (Lespinasse, *Métiers et corporations*, vol. 3, p. 547) but in all likelihood they were not the whole guild. In the 1590s an Italian visitor recorded that there were twenty-two clockmakers' shops in Paris, cf. Raynaud, *Paris*, p. 166.

43. The new regulations of the guild ratified in 1646 established that the number of master clockmakers in Paris should be limited to seventy-two (cf. Lespinasse, *Métiers et corporations*, vol. 3, p. 555).

44. Develle, *Horlogers Blésois*, pp. 20–21.

45. Vial and Côte, *Horlogers Lyonnais*, pp. 11 and 22.

sixteenth century[46] and their number grew very rapidly in the following decades. In Augsburg in 1615, there were forty-three master clockmakers beside forty-three journeymen who had come to Augsburg from other towns and had not yet been granted citizenship.[47] In London there were probably more than sixty masters in the 1620s.[48] In some of the centres where they grew in numbers and in wealth, the clockmakers eventually established their own guild – in Paris in 1544, in Blois in 1597, in Geneva in 1601, in Toulouse in 1608, in London in 1631, in Lyons in 1658–60, in The Hague in 1688, in Stockholm in 1695, in Copenhagen in 1755. These cases should not be generalized. In many a town, the clockmakers never had their own guild and instead belonged to the guild of the smiths. In many other towns there was scarcely a clockmaker until very recent times. Interestingly enough, Italy, which had been a leading country in matters of horology during the Middle Ages, did not develop any great centre of clock and watch-making in the modern age. Outstanding clockmakers were not completely lacking and skilful craftsmen were often invited from France and Germany by Italian princes. But in the course of the sixteenth and seventeenth centuries the country underwent a drastic process of social involution which was concomitant in the seventeenth century with a process of commercial and industrial decline. Thus the situation that eventually prevailed in Italy was one of a few outstanding craftsmen who worked mostly under the patronage of Princes, Popes and other dignitaries producing few artistic pieces for aristocratic circles.

As long as most of the clocks were huge public time-keepers made of iron or bronze it is perfectly understandable that their makers were smiths, locksmiths, gunsmiths, and metal workers in general, but the situation changed in the course of the sixteenth and seventeenth centuries when domestic clocks and watches became less of a rarity. The smaller timepieces were expensive

46. Babel, *Histoire corporative de l'horlogerie*, pp. 390*ff*.

47. According to a census of the population of Augsburg in 1610, there were forty clockmakers in the town. Another census of 1615 gives the names of forty-three clockmakers and indication of forty-three journeymen (not citizens). Another census of 1619 gives again the number of forty-three clockmakers. I owe this information to the kindness of Dr H. F. Deininger, Director of the *Staatsarchiv* in Augsburg.

48. Cf. below note 89 to p. 142.

contrivances and were owned by well-to-do people. Being luxury items they were at the very centre of the craze for exuberant decoration that characterized the late Renaissance and the Baroque Age. The craftsmen who had to satisfy the new vogue now needed the skills of the goldsmiths rather than those of the blacksmiths and locksmiths. In England, France, Germany, Italy and many other places a clear-cut distinction developed between 'makers of big public clocks' and 'makers of little clocks and watches'. The materials used by the 'makers of little clocks and watches' were often expensive and for those who specialized in the production of highly ornate clocks and watches the necessary investment in working capital was quite substantial. Masters with a rich clientele or princely patrons enjoyed good economic conditions. In those centres like Geneva where, as we shall see, clockmaking developed into a prominent industry, clockmakers also acquired a prominent social position. In general, however, clockmakers were not wealthy people and clockmaking was not considered an especially rewarding profession. A French document of the 1580s ranking for fiscal purposes the professions practised in Paris in five classes according to their estimated economic worth, put clockmaking in the third class.[49] In the Netherlands, in the late seventeenth and early eighteenth centuries, professions were ranked in four classes and clock- and watchmaking was put in the second class.[50] These broad generalizations probably did not err on the side of optimism.[51]

Who were the clockmakers and what was their social origin? John de' Dondi, the scholar, was a special case, although not a rare one, and the same can be said for the friars who turned clockmakers. The great majority of clockmakers came from the large social stratum of artisans that was a characteristic of European towns. Of thirty-three clockmakers who worked in Lyons in the period from 1550 to 1650 (and whose fathers'

49. Lespinasse, *Métiers et corporations*, vol. 1, pp. 94–6.

50. Oldewelt, *De beroepsstructuur van de bevolking der Hollandse*, pp. 81, 82 and 83.

51. Taxes actually paid by watchmakers in Rotterdam and Leiden in 1674 seem to indicate that these craftsmen were generally in the lowest income groups, and not in the second from the top as indicated by the general classification mentioned above in the text. Cf. Oldewelt, *De beroepsstructuur van de bevolking der Hollandse*, pp. 80*ff*.

professions we know) thirteen were sons of clockmakers and two each were sons of goldsmiths, merchants, teachers and tailors. Of the remaining twelve, one was the son of an apothecary, one of a surgeon, one of a shoemaker, one of a metal-founder, one of a gunsmith, one of a mint master, one of a lace worker, one of a proxy, one of a carpenter, one of a clerk, one of a constable, and one of an unskilled labourer. In Blois (France), among clockmakers and apprentices who worked there from 1550 to 1700, more than sixty-five were sons of clockmakers, four were sons of merchants, three each were sons of goldsmiths and carpenters, two were sons of surgeons, one was the son of an apothecary, one of a clerk, one of a notary, one of an architect, one of a gunsmith, one of an armourer, one of a tailor, one of a locksmith, one of a constable and one of a tax collector.[52] Other examples that can be collected from other areas confirm this picture. An apprentice clockmaker in Geneva in 1569 was the son of a gunsmith, another in 1672 was the son of a professor of philosophy and a third one in 1674 was the son of a doctor.[53] In England Thomas Tompion was the son of a blacksmith, John Harrison was the son of a carpenter, Thomas Mudge was the son of a clergyman and John Arnold was the son of a watchmaker.[54] Jacob Deburges, who was born in England and moved to Blois in the 1590s, was the son of a bookseller.[55] In seventeenth-century Basel two apprentices were sons of clockmakers and one was the son of a deacon.[56] In those places where guilds of clockmakers were established, the corporative system obviously influenced the social structure of these local groups of clockmakers. For one thing, it emphasized and institutionalized the distinction between masters and labourers; for another, with the establishment of

52. For Lyons cf. Vial and Côte, *Horlogers Lyonnais*, pp. 29–97; for Blois cf. Develle, *Horlogers Blésois*, pp. 214–450. The high percentage of clockmakers, sons of clockmakers, indicated above in the text is biased and inflated. The reason is that those authors who wrote about the history of horology knew the profession of the father of a clockmaker especially if he too had been a clockmaker. From the biographies compiled by Vial and Côte for Lyons, and Develle for Blois, one also gathers that master-clockmakers very often married daughters of clockmakers.

53. Babel, *Histoire corporative de l'horlogerie*, pp. 68, 418, 419.

54. Britten, *Old Clocks*, pp. 276, 286, 288.

55. Develle, *Horlogers Blésois*, p. 262.

56. Fallet-Scheurer, *Uhrmacherkunst in Basel*, pp. 152–3.

high and differentiated fees for apprenticeship or for acceptance into the guild, it restricted the area from which the masters could be drawn and it strengthened the tendency to make the profession hereditary within certain families.[57]

As a rule, medieval scholars were not interested in machines, but the clock, because of its connexion with astronomy, was an exception. Mention has already been made of the contributions to horology by medieval friars and by the two Dondis, both '*philosophie, medicine et astrologie doctores*' at the University of Padua. In the Renaissance, while clocks and watches were becoming progressively more fashionable in the upper classes as useful and graceful ornaments, the clock as a machine attracted more and more the inquisitive curiosity of scholars, amateur scholars and learned people in general. According to Vasari, Brunelleschi 'together with some studious persons, looked into the matter of time and of movements, of weights and wheels, and how to make them work and he himself made some very good and very beautiful clocks'.[58] Lorenzo della Volpaia was '*eccellentissimo maestro d'oriuoli et ottimo astrologo*' and built a very elaborate public clock for Lorenzo il Magnifico.[59] The great Leonardo was interested in clockwork and he actually depicted in one of his sketches a kind of escapement that seems to indicate that he had some idea of using the pendulum as a time-regulator. In the sixteenth century two famous mathematicians Chretien Herlin and Conrad Dasypodius worked at the great clock of Strasbourg Cathedral.[60]

In the seventeenth century, when the Scientific Revolution exploded in all its exuberance and vigour, the champions of the new science manifested an avid interest in horological matters. In their eyes the clock was the machine *par excellence* and it fascinated them. But there was more to it than that. The sixteenth and seventeenth centuries witnessed the great astronomical discoveries and the great expansion of ocean navigation: both astronomers and navigators needed precision time-keepers to

57. In the guilds the admission fee was lower for the sons of clockmakers than for other applicants.

58. Vasari, *Vite*, vol. 1, p. 593. Cf. also above note 24 to p. 123.

59. Morpurgo, *Orologiai della Volpaia*, pp. 23–6; Morpurgo, *Dizionario*, pp. 201–4.

60. Ungerer, *Horloge astronomique de Strasbourg*, pp. 14*ff*.

determine longitude and the right ascension of the stars. At the same time, the construction of high-precision time-keepers presupposed the solution of basic problems of mechanics that were at the very heart of the Scientific Revolution. Among those who devoted themselves to the problems of measuring time and of constructing accurate clocks one can mention Galileo, Christian Huygens, Robert Hooke, Godefroy Wendelin, Nicolas Fatio and Wilhelm Leibniz.

Until the middle of the sixteenth century the verge escapement with foliot remained unchallenged and the construction of all time-keepers depended on that device. In the course of the seventeenth century, however, when scientists devoted their attention to the problem of the measurement of time, they applied scientific principles and systematic experimentation to horology. Scientists and clockmakers then worked in close collaboration and the result was a series of revolutionary discoveries and a break-through in the technological progress of the craft. The most important step was the introduction of a new mechanism which used the pendulum instead of the foliot as the controlling device. Galileo had had in mind a solution of this sort, but it was Christian Huygens who solved the problem in the late 1650s and inaugurated the production of pendulum clocks.

The importance of the innovation is shown in the graph on page 134. For the period preceding 1800, the values represented in the graph are of course only rough estimates of the average accuracy of the best clocks of the time, but with all their limitations the data are sufficiently revealing to indicate the importance of Huygens' discoveries. The appearance of the pendulum ended the era of the low-precision time-keepers and opened the era of the high-precision instruments. On a more general level, the graph shows the dramatic discontinuity in the history of technology that took place in the course of the seventeenth century, in connection with the so called 'Scientific Revolution'. The progress in applied technology that followed that 'Revolution' appears to be of exponential kind. In the narrow field of horology a series of important discoveries and innovations accompanied and followed the birth of the pendulum clock, but all that was happening in horology was but one aspect of a broad and complex

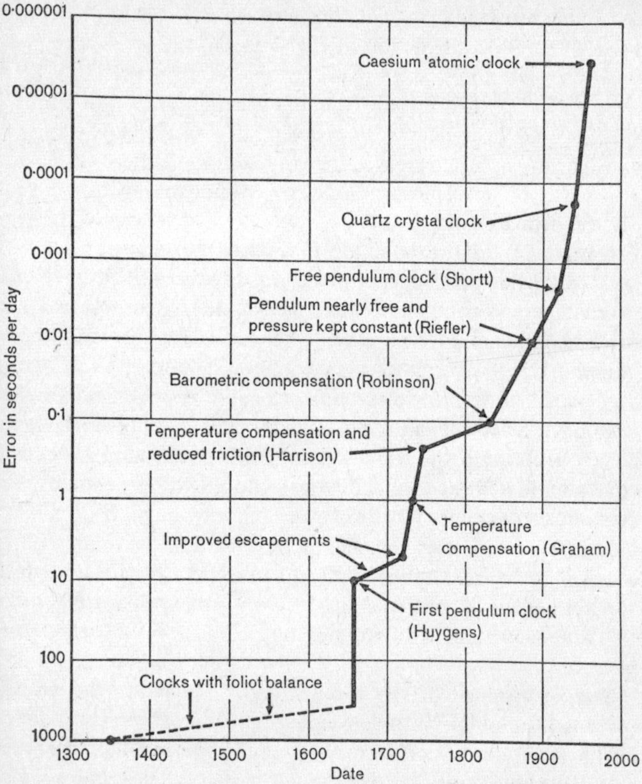

set of interlocking technological changes. Clockmaking was the first industry to put into practice the theoretical findings of physics and mechanics. At the same time, it set the pace for the general development of applied mechanics and played a role of prime importance in the evolution of scientific instruments. As has been written,

the earliest achievements in precision mechanics, and the essential steps in its progress up to the present day, must be credited to horology.... At an early date clockmakers devised tools with which the most delicate operations of their trade could be accurately performed ... they were led to investigate the properties of the types of copper and steel used in their work, to study the thermal expansion of metals, the elasticity and resilience of springs. They invented and perfected machines for the manufacture of some of their simpler tools.

Thus, apart from strictly horological inventions, thanks to which the precision of timepieces was continually improved, the clockmakers placed at the service of the science of mechanics an array of equipment which was constantly being perfected and had only to be taken over, directly or indirectly, by the constructors of instruments.[61]

It was not by chance that the century that witnessed the appearance of the first accurate timepieces was also the century that witnessed the birth of the industry of scientific precision instruments in general. '*Les instruments précis font progresser la science, tandis que la science permet l'amelioration des instruments précis*': after a long period of gestation, our modern scientific age was born – and cumulative processes progressively accelerated its further development. Lewis Mumford overstates the case when he writes that 'the clock, not the steam engine, is the key machine of the modern industrial world', but his hyperbolism contains more than a grain of truth.

In 1745, referring to the export of English clocks, J. Cary wrote that 'we sell little but art and labour, the material whereof clocks are made being but of small value'.[62] In 1843, in an evaluation of costs in the production of watches in Geneva, J. A. Bloch Borel estimated that the raw materials (gold and silver) accounted for about two-sevenths of the cost and labour for about five-sevenths.[63] It may well be that in the sixteenth and seventeenth centuries on the average the ratio was somewhat more favourable to the raw materials input, but even for those centuries one can hardly over-estimate the relative importance of labour. It must also be noted that the raw materials used in the construction of chamber clocks and watches were of light weight and small volume and could easily be transported over long distances at a reasonable relative cost, even in an age when transport was far from easy. Thus the availability of raw materials played only a small role in the geographical location of the industry. Two other factors seem to have been far more im-

61. Daumas, *Instruments scientifiques*, pp. 155*ff*.
62. Cary, *Discourse on Trade*, p. 21.
63. Babel, *Histoire corporative de l'horlogerie*, p. 38, *n*. 1.

portant: the demand for timepieces and the supply of skilled labour. The growth of the clockmaking industry in Blois and Paris was undeniably related to the demand of the court, the aristocrats, and the rich bourgeois for clocks and watches. In other places it was a combination of local demand and commercial organization that captured foreign demand and transmitted its stimuli. Demand was a necessary factor but it alone was not sufficient to account for the growth of clockmaking in certain areas; the supply of skilled labour must also be taken into consideration.

I have already mentioned that there were no centres of clock manufacturing worthy of the name until the end of the fifteenth century. Toward the end of the century, however, Augsburg and Nuremberg developed into such centres. In the course of the Middle Ages both towns had developed a great tradition in metal-working. Moreover, the two towns were great commercial centres that offered good opportunities for exports. The combination of these circumstances allowed clockmaking to grow and prosper.

By the beginning of the sixteenth century, the Nuremberg and Augsburg masters and their products were known all over Europe. Their fame became so great that Peter Henlein of Nuremberg was for a long time incorrectly credited with the invention of the watch. We have no information about the number of clockmakers in Nuremberg. For Augsburg three censuses of 1610, 1615 and 1619 indicate that there were about forty master clockmakers and an almost equally large number of workmen in the town.[64]

The period of brilliant economic growth seems to have come to an end toward the middle of the sixteenth century but in matters of horology the fame of the two German towns lasted well into the first quarter of the seventeenth century. Still at the end of the sixteenth century, timepieces made in Augsburg and Nuremberg were being exported to distant foreign markets, and the Italian Garzoni wrote that 'a great number of German craftsmen excel in the art of horology and all the best and the most precise timekeepers actually come from Germany'.[65] Circa 1600

64. Cf. above, note 47 to p. 129.
65. Garzoni, *Piazza Universale*, p. 625.

Fynes Moryson wrote that 'tuching manuall arts, the Dutch are a people more industrious then the Germans and excell them in all arts and trades . . . howsoever I must confesse that the Germans of Nurenberg in those parts are esteemed the best workmen for clockes and some like thinges.'[66] This supremacy was brought to an end by the decline of some of Nuremberg's and Augsburg's traditional customers such as Italy in the South and Antwerp in the North as well as by the havoc of the Thirty Years' War (1618–48). In Augsburg, according to a census, only seven master clockmakers were left in town in 1645.[67]

Clockmakers had many traits in common with the other craftsmen of their age: with few exceptions they were violent and riotous in their youth, litigious in their adult years, and dishonest whenever the occasion arose. Among them there were, of course, the illiterate as well as the literate ones, but on the whole clockmakers were distinguished among other craftsmen for a relatively high degree of literacy.[68] In point of fact, a number of clockmakers were educated far beyond the level of reading and writing. Julien Coudray and Guillaume Coudray, clockmakers of Blois in the second half of the sixteenth century, frequented the Royal Library. Christopher Piron, who died in Blois in 1637, left in his estate '*une quantité de livres d'astrologie*' and Jacques Duduict, who worked there in the early 1600s, was a good writer. Master Jehan Flanc, '*horlogeur de la Ville*', who died in La Rochelle in 1616, was '*sçavant aux mathématiques et aux fortifications*'.[69] Clock-making required certain skills in drawing and it was essential for clockmakers to be able to count.[70] In a number of contracts of apprenticeship made in Geneva in the course of the seventeenth century it was expressly stated that the master had

66. Moryson, *Itinerary*, p. 372.

67. The information was kindly given to me by Dr Deininger.

68. For Lyons cf. Vial and Côte, *Horlogers Lyonnais*, pp. 20–21. In London, among forty-eight masters who signed a document of the guild in 1632, only three signed with a mark; cf. Atkins and Overall, *Company of Clockmakers*, p. 51. Out of nine clockmakers who married in Bristol between 1661 and 1694 only one signed with a mark; cf. Ralph, *Marriage Bonds*, and Firth, *Marriage Allegations*.

69. Develle, *Horlogers Blésois*, pp. 214, 224, 267, 287; Bourriau, *Horlogers à la Rochelle*, pp. 15–16.

70. Of course illiterate master-clockmakers were not totally absent: cf. Vial and Côte, *Horlogers Lyonnais*, p. 20, and above, note 68.

'to teach the apprentice how to read and write, and this especially in the first year of the apprenticeship'.[71]

The Reformation apparently made a relatively large number of converts among the clockmakers. Of ninety master clockmakers who lived in Catholic Lyons during the period 1550–1650, and whose religion we know, fifty were Catholic and forty were Protestant. We had no precise figures for Paris but some evidence seems to indicate that the situation there was similar to the one existing in Lyons. Information about clockmakers who turned Protestant is also available for other places such as Blois, and La Rochelle.[72] It is reasonable to suppose that there was some connexion between the high degree of literacy prevailing among clockmakers and their relatively high propensity to follow the Reformation movement. Whatever the reasons for their religious inclinations, the clockmakers were often persecuted in the southern Low Countries, France and Italy, and thus one encounters them frequently in the history of those migrations of skilled labour that helped to change the balance of economic power in Europe.

Here in fact lies one of the fascinating aspects of the social history of horology in those days: until at least the end of the seventeenth century the industry did not need great capital equipment or an elaborate organization; given satisfactory conditions on the side of demand, a handful of craftsmen was all that was needed to make a great centre of production. As I have already indicated, there were less than two dozen master clockmakers in Paris in the middle of the sixteenth century, and the masters who made Lyons a centre of the clock industry were no more numerous. To destroy or to build up the industry it was enough to dismiss or to attract a dozen craftsmen. We, today, are not so vulnerable at the level of horology, but at higher levels the selective massacre of a dozen scientists would bring more than one branch of science to a standstill. While Nuremberg and Augsburg declined, Geneva

71. Babel, *Histoire corporative de l'horlogerie*, p. 419.

72. For Lyons cf. Vial and Côte, *Horlogers Lyonnais*, p. 20; for Paris cf. Franklin, *Vie privée*, p. 144; for Blois, cf. Develle, *Horlogers Blésois*, pp. 426 and 447. La Rochelle was one of the main centres of the Reformation movement in France and significantly enough printing and clockmaking flourished there in the seventeenth century (Bourriau, *Horlogers à La Rochelle*, p. 6).

and London emerged as major centres of clockmaking. In both places the development of the industry was related to the inflow of a handful of refugees[73] – to the injection of a small but precious amount of human skills.

In 1449 there was one clockmaker in Geneva[74]. In the tax-list of 1464 no clockmaker is mentioned.[75] In 1515 when the clock of the church of St Pierre got out of order, there was no craftsman in town who could repair it. Shortly after 1550, a few clockmakers began to arrive in Geneva as refugees: Philippe Bon and Master Bayard came from Lorraine, Master Somellier from Dieppe, Laurent Drondelle from Paris, Pierre Charpentier from Orleans, Pierre de Fobier from Languedoc, Evangeliste de Marco from Foligno (Italy), Denis de Saint Paul and Clement Bergier from Lyons, Nicolas Bissac and Charles Cusin from Burgundy. Geneva's gates at that time were open to all who were escaping from religious persecution; it was actually in Calvin's interest to increase the number of his adherents after he had defeated the party of the Libertines. The immigration does not seem to have caused economic difficulties. On the contrary, the rapid growth of the working population was accompanied by a remarkable economic renaissance, partly because the immigrants were mostly skilled craftsmen and partly because Geneva was traditionally a commercial centre with fairs that offered excellent opportunities for exporting. The ensuing prosperity also brought to Geneva people who, as a contemporary document disapprovingly mentions, were attracted more by the opportunity of 'practising their craft than by the Gospel'. Clockmakers were not very numerous among the refugees[76] but as I have said before a handful of craftsmen was all that was needed in those days to make a centre of clock manufacture; on the other hand, those few clockmakers proliferated rapidly. About 1600 there were in Geneva twenty-five to thirty master clockmakers and an unknown number of work-

73. Other centres of clockmaking developed in the Zaanstreek and in Friesland. Also in these places the French refugees seem to have played an important role in the development of the clock manufactures. Cf. Ottema, *Uurwerkmakerskunst in Friesland*, p. 9.

74. Babel, *Histoire de Genève*, vol. 2, p. 109.

75. Bergier and Solari, *Histoire*, pp. 203–9 and 214–24.

76. Geisendorf, *Métiers et conditions sociales du premier Refugé*, p. 247, n. 2.

men and apprentices.[77] By the 1680s there were, according to Gregorio Leti, more than one hundred masters and about three hundred workers who produced more than 5,000 timepieces a year.[78]

London's story is not unlike that of Geneva. During the reigns of Henry VIII and Elizabeth I, a number of foreigners were allowed to settle in England. Often the immigrants aroused the hostility of English craftsmen and the May Day riot of the London Apprentices in 1517 against the 'Stranger Artificers' shows to what degree of frustration the English craftsmen were driven by alien competition. At times also the legislation seems to have been calculated to deter rather than to encourage the advent of foreign artisans. However, the aliens were successfully drawn into the industrial system of the country. On the other hand, the reaction of English craftsmen should not be misinterpreted. It did not signify lack of receptiveness to foreign ideas and technologies: it sprang from a widespread jealousy of foreign artisans. 'English,' wrote Fynes Moryson, 'prefer strangers as well phisitians as other like professors then their owne countrymen as more learned and skillful then they are, which makes the English also so much travayle in forrayne parts and so much esteeme theire owne countrymen being travelers.'[79]

In *Hedda Gabler*, when Ibsen wanted to show Hedda's husband as a dull and pedantic man, he mentioned that the husband was at work on a study of the textile industry in the sixteenth century. My opinion of scholars concerned with the sixteenth century textile industry is a more charitable one, but I suspect that in recent times excessive concern with the stagnation of export of English woollens after 1550 has dulled our view of English economic and social history in Elizabethan times. Some scholarly circles have devoted much time and brilliant energies to prove that there was no Industrial Revolution before the Industrial Revolution. One can easily agree with this unexhilarating view, yet one can still submit that in Elizabethan England innovation was the strikingly predominant note in many basic fields such as

77. For all that precedes and follows cf. Babel, *Histoire corporative de l'horlogerie*, pp. 43–88 and 391–3.

78. Leti, *Historia ginevrina*, part 4, pp. 612–13.

79. Moryson, *Itinerary*, p. 475.

education, shipbuilding, navigation, overseas trade, the building trades and the iron industry not to mention the 'quantitative' approach by which some of these activities were being analysed for the first time.

In matters of horology English backwardness was conspicuous until the last decades of the sixteenth century. When Henry VIII wanted to have some work done on the clock at Nonsuch Palace he had to import French clockmakers to do it.[80] Nicholas Cratzer 'deviser of the King's horloges' was a Bavarian.[81] As for the production of watches, there is no record of an English watch before 1580 and only a short time before this date is there a record of a watchmaker.[82] But before the century was over signs of improvement and change were abundantly clear. The demand for timepieces was growing and foreign craftsmen were flowing into the country. Nicholas Urseau, clockmaker to Queen Elizabeth, was of French origin.[83] François Nawe who was producing clocks and watches in London in the 1580s[84] must also have been French to judge from his name. Their example was not lost, because Elizabethan Englishmen were conspicuous for some of the qualities that we nowadays associate with the Japanese. The first watches made by Englishmen in the last two decades of the sixteenth century were unimaginative but diligent replicas of French and German models. A watch made by Bartholomew Newsam has a pierced cover of the German type. The striking watch made by Randolf Bull of London (now in the Mallet collection of the Ashmolean Museum at Oxford) is a good example of the composite type. The mechanism is French, the case is German, and the inner ring of hours from 13 to 24 on the dial is suitable for a system of counting time which was used in Germany but not in England.[85] In some cases English makers used for their watch and clock movements continental-made cases, which displayed a far finer quality and richness of design than any of English make.[86]

80. Cunningham, *Alien Immigrants*, p. 215.
81. Page, *County of Middlesex*, p. 158; Symonds, *English Clocks*, pp. 26–7.
82. Britten, *Old Clocks*, p. 43.
83. Page, *County of Middlesex*, p. 158.
84. Britten, *Old Clocks*, p. 45.
85. ibid.
86. Symonds, *English Clocks*, p. 30.

Although they were willing to imitate foreign products, English clockmakers were not at all happy about the unremitting immigration of foreign artisans. In 1622 the 'Clockmakers Citizens and inhabitants in London' angrily complained to the Crown that they were 'interrupted and discredited in the use of their trade by the interposition of manie straingers invadinge this Realme', and of course they accused the 'straingers' of all sorts of 'abuses'.[87] Five years later the 'Free Clockmakers of the City' complained again of being 'exceedingly oppressed by intrusion of clockmakers strangers' and especially by 'the Ffrench clockmakers'.[88] According to a list of names attached to the document of 1622 there were sixteen English clockmakers 'householders' in London and about thirty 'knowne straingers of ye same arte dwelling in and about London' as well as a certain number of foreign apprentices. The names of the 'straingers of ye same arte' indicate in most cases a French origin. Although most of the French masters were Huguenots, a few were not. In *A True Certificat of the Names of the Straungers residing and dwellinge within the City of London* taken by the Privy Council in 1618, one finds in the ward of Farrington within, a 'Bernaby Martinot, clock-maker, born in Paris, a Roman Catholique' and in Portsoken Ward a 'John Goddard, clock-maker ... a papist'.[89] Incidentally, in the list of 1622 one finds a Lewis Cuper who had emigrated to London sometime between 1613 and 1622. The Cupers were at that time a famous family of clockmakers in Blois, but the family was of German extraction and had settled in Blois sometime before the middle of the sixteenth century: in this roundabout way skills and technologies were broadcast throughout Europe.

There are reasons to suspect that the list of 1622 maliciously underestimated the number of English clockmakers, but even if we were to double the number declared, the resulting ratio between 'English' clockmakers and 'straingers' would still prove that in London as in Geneva the most relevant factor in the growth and development of the clock- and watchmaking industry was the immigration of foreign craftsmen.

87. *Calendar State Papers*, Domestic, James I, 127/15 and 16, January 1622.
88. Atkins and Overall, *Company of Clockmakers*, p. 2.
89. Ullyett, *British Clocks*, p. 18.

As late as 1657 England was still borrowing from the Continent in horological matters: in that year John Fromanteel, a member of a family of Dutch extraction, went to Holland to learn the art of making pendulum clocks of the type recently invented by Huygens and made by Coster; on John's return the Fromanteels were the first to make pendulum clocks in England.[90] However, in the course of the seventeenth century English makers made noticeable progress, establishing original lines of thought and acquiring an undisputed supremacy over their continental colleagues. With men of inventive genius such as the famous Dr Hooke (1635–1703), they improved the precision of timekeepers and invented a number of ingenious horological devices. Among such devices one should mention the anchor escapement that appeared around 1670. This escapement had the advantage of greatly reduced wear. Furthermore it enabled the use of a long pendulum with a small arc without recourse to intermediate gearing, thus allowing for more precise measurement of time. One should also mention the repeating mechanism. In an age when the lighting of a house was not a simple matter, telling time was a difficult problem at night. In France M. de Villayer allegedly solved the problem by designing a clock with a dial that had different kinds of spices inserted in the place of numbers. At night he reached for the point indicated by the hour hand and tastefully determined the time. English clockmakers searched for less gastronomical solutions and in the 1680s Daniel Quare and Edward Barlow produced repeating devices for watches and clocks.[91]

By 1680 England had attained an unchallenged preeminence in the field of horology. At the same time an unmistakably English style emerged in the form and decoration of case work.

By the dawn of the eighteenth century, London and Geneva were by far the two greatest centres of clock- and watchmaking in Europe. With the emergence of these two centres, new proto-industrial and proto-capitalistic ways of doing things appeared in the trade and in the making of timepieces. Especially after the first half of the seventeenth century, specialization developed among craftsmen in the more advanced clockmaking centres with regard to the production of single parts of timepieces. The

90. Britten, *Old Clocks*, p. 272.
91. ibid., p. 280.

spring-makers were apparently the first specialized workers to appear, but other specialists soon followed: at the beginning of the eighteenth century many streets in Clerkenwell, a district of London, were occupied by workmen who were escapement-makers, engine-turners, fusee-cutters, secret-springers and finishers. As early as 1701 the making of a watch was chosen to illustrate the advantages of division of labour. In Geneva two groups of specialized workers succeeded in having a guild of their own, distinct from the guild of clockmakers: thus the assemblers had their guild recognized in 1698 and the chiselers in 1716. Quite naturally these developments affected trade also: one reads only of clocks and watches being traded and sold before 1600, but after that date there are frequent references to transactions concerning single parts of clockwork. At the same time, the changed conditions caused the actual craftsman to become divorced from the retail side of the business.

The corporative structure with its distinction among masters, journeymen and apprentices remained in existence but above these three categories a new group emerged, numerically small but economically powerful. This was the group of the clock merchants, proto-capitalistic entrepreneurs who placed orders with the masters, advanced funds to them, and arranged for the marketing of the merchandise at home or abroad. This particular type of organization seems to have developed more fully in Geneva.

These developments came to full maturity in the course of the eighteenth century and resulted in the mass production of common and relatively inexpensive timepieces. Savary noted that Geneva exported 'some excellent timepieces, but a much greater quantity of common ones'[92] and Adam Smith wrote that:

> the diminution of price has, in the course of the present [eighteenth] and preceding century, been most remarkable in those manufactures of which the materials are the coarser metals. A better movement of a watch, that about the middle of the last [seventeenth] century could have been bought for twenty pounds, may now perhaps be had for twenty shillings.[93]

At the same time technological progress made possible further

92. Savary, *Dictionnaire*, vol. 3, col. 342.
93. Smith, *Wealth of Nations*, p. 243.

substantial improvements in the precision of timepieces (see the graph on p. 134). The reverse of the coin, the price paid for greater efficiency on both the economic and technological planes, was more vulgarity and cheapness on the artistic level. The mass-produced timepiece with interchangeable parts, and made by specialized workers, was leading the way toward the Industrial Revolution.

Much of what was happening on the side of production was interlocked with what was happening on the side of demand, namely the growth of a steadily rising proportion of middle-class and relatively prosperous people who could afford to buy time-pieces. The combination of developments both on the side of demand and on the side of supply resulted in a greater diffusion of clocks and watches and Savary could write that 'horology ... has become one of the most active and advantageous branches of the trade'.[94] It is not surprising therefore that in a number of countries more attention was devoted to this 'branch of trade'.

The idea of developing a clock and watch industry appealed to the tastes and inclinations of the enlightened rulers as well as of the enlightened philosophers and scientists of the eighteenth century. In the second half of the century the King of Prussia repeatedly tried to establish a manufacture of timepieces in Berlin. A similar attempt was made by the Margrave of Baden in 1767. A pilot factory for the production of watches and clocks was established at Dubrovno in Bielo-Russia in 1784, and in 1792 it was moved to Kupavno, near Moscow. About the same time the King of Portugal established a clock factory in Lisbon. The engineer and inventor Polhem had a factory operating at Stjernsund, in Dalecarlia (Sweden) at the beginning of the eighteenth century; and Voltaire set up a manufacture at Ferney (France) in the 1770s. These and other initiatives met with different fates; a few of them succeeded, most of them failed. The most interesting and lasting developments anyhow occurred in Paris.

As has been indicated in previous paragraphs, Paris was one of the first towns to emerge in Europe as a centre of the clock-making industry. Political troubles and religious strife possibly hampered the development of the industry in the second half of the sixteenth century, but although we lack precise information,

94. Savary, op cit., vol. 3, col. 329.

there are reasons to believe that in the first half of the seventeenth century the industry was still expanding. In 1646 the clockmakers felt the need to limit to a maximum of 72 the number of masters admitted to the guild[95] and this decision may reasonably be assumed to indicate that the number of clockmakers in Paris was growing. From the middle of the seventeenth century, however, the situation deteriorated. The guild regulations hampered development and opposed innovation while the competition of Geneva grew progressively stronger. Amidst these troubles, the Edict of Nantes was repealed. There is no doubt that the economic effects of the Revocation have been occasionally exaggerated, but it would be difficult to maintain that what happened in 1685 and in the following years did not damage the clockmaking industry. Professor Scoville has proved that the Huguenots who left France were a very small proportion of the total active population and he deduces from this that the consequences of the emigration cannot have been of great relevance.[96] The argument is sound as far as it goes but it does not go very far: it may hold for those branches of the economy in which large numbers of craftsmen were available, but horology was a field in which, as we have already seen, a small number of craftsmen could make quite a difference. Whatever their relative weight, the combination of restrictive guild regimentation, foreign competition and emigration of rare and skilled human resources proved highly disruptive. At the beginning of the eighteenth century, an official paper reported on the bad conditions of the French clock and watch industry and pointed out that it was impossible to have a timepiece in France that did not contain parts imported from London or Geneva.[97] It was at that juncture that the French administration took steps to improve the conditions of the industry.

History works often through curious ways. As has been indicated above, London largely owed the development of her horological industry to the immigration of French craftsmen. At the beginning of the eighteenth century, in order to revitalize the French industry, the French administration invited to Paris an

95. Lespinasse, *Métiers et corporations*, vol. 3, p. 555.

96. Scoville, *The Persecution of Huguenots, passim* and especially pp. 434–47.

97. Franklin, *Mésure du Temps*, p. 144.

English clockmaker of great repute, Henry Sully, and with him sixty other English craftsmen. The establishments set up by Sully at Versailles and at St Germain did not take firm root and disappeared within a few years; but, as the French themselves recognized, although the venture proved a failure, it nevertheless introduced to French masters a standard of workmanship superior to anything they had previously known. The products of Sully '*excitèrent l'émulation parmi les Horlogers de Paris*' and led to great advances in French horology, principally under the leadership of Julien Le Roy.

The progress of the industry was not an easy one: it was harassed by the unremitting competition of the Swiss manufactures. Low cost products from Geneva flooded the French market and there was nothing that the Swiss would not do to push their products. The Swiss clockmakers did not sign their clockwork so that on occasion they could sell them to other manufacturers who would then mark them with their own signature; if London-made parts were requested, then the Swiss did not hesitate to imprint the mark of London on their own products. In Paris, according to the brothers Castel, clockmakers in Bourg-en-Bresse, through some kind of bribery the Swiss managed 'to keep stocks of watches and clockwork in the Abbey of St Germain and other privileged places supplying many a clockmaker with their products'. 'The Swiss watch industry ruins our manufactures', sadly remarked Beliard just after the middle of the eighteenth century and the brothers Castel added that 'because of this trade of timepieces from Geneva a considerable amount of money flows out of this kingdom every year'.

Yet these and other difficulties did not prevent the recovery of the French manufactures. The economic history of France in the two centuries 1550 to 1750 is an intriguing puzzle to the economic and social historian. To the dispassionate observer the country seems always on the verge of a final collapse and all the odds seem to be against any possible recovery. Yet as the legendary phoenix, consumed in fire by her own acts, France always managed to rise in youthful freshness from her own ashes. An incredible amount of robust human energies must have been constantly fermenting under the troubled surface. In the narrow field of horology there is no doubt that after the first quarter of the eighteenth century the industry grew steadily and before the century was over, Paris was with London and Geneva one of the three main centres of the clock and watch industry in the world. By the 1770s an Italian author wrote that 'English craftsmen acquired great repute in Europe for the beauty and the precision of their products . . . but nowadays the clockmakers of Paris excel the English for the beauty and the neatness of their work.'[98] In the 1780s there were in Paris possibly more than 400 master clockmakers.[99] French watches were sold in London. However, until the end of the century, a more efficient commercial organization greatly favoured the products of London and Geneva, which largely prevailed on the international market. Toward the end of the eighteenth century London produced a yearly average of about eighty thousand timepieces for export and about fifty thousand timepieces for the home trade.[100] Geneva possibly produced an average of seventy or eighty thousand pieces, which were, of course, almost all exported.[101] London and Geneva timepieces were then popular even in Turkey and exotic, distant China.

98. Griselini, *Dizionario*, vol. XI, p. 4.

99. Reverchon, *Histoire de l'horlogerie*, p. 151.

100. Atkins and Overall, *Company of Clockmakers*, p. 263.

101. Chapuisat, *Commerce et industrie à Genève*, p. 217, who derives his data from Sismondi. A French report written by two clockmakers brings the figure to 250,000 watches, but it was an obvious exaggeration (Babel, *Histoire corporative de l'horlogerie*, p. 398. Cf. also Babel, *Fabrique genevoise*, pp. 51–2, and Say, *Économie Politique*, vol. 1, p. 187).

Chapter 2

Chinese Mandarins and the Self-Ringing Bell

'A cynic might observe that if it was Christians and spices which had brought the Portuguese to the Orient, it was the spices which were instrumental in keeping them there.'[1] This remark by Professor C. R. Boxer is applicable as well to the motivations that brought and kept overseas the Spaniards, the Dutch, the English, the Danes and the Swedes. Missionaries apart, Europeans went to the East solely to trade and until the end of the eighteenth century they were sensible enough to realize that overseas territorial conquests were not within the realm of reasonable possibility. Whatever conquests they attempted were, with a few exceptions, limited to islands and harbours to be used as entrepôts for their trading activities. Knowing that their technological and military superiority resided in their gun-carrying, ocean-going sailing ships, the Europeans were satisfied for almost three centuries to limit their control to the high seas.

When they first reached the Far East, Europeans were chiefly interested in spices but it did not take them very long to realize that a number of other commodities offered excellent opportunities for profit, e.g., copper from Japan, cotton textiles from India, silk and carpets from Persia, silk and porcelain and (with the end of the seventeenth century) tea from China. The chief obstacle to trade, however, was the fact that they had almost no goods to offer in exchange for eastern products. Since the Industrial Revolution, we have come to take western superiority in the arts and sciences so much for granted that it is difficult for us to imagine a situation in which the East had much to offer in raw materials

1. Boxer, *Portuguese in the East*, pp. 192 and 214.

and manufactured goods while the West had very little that appealed to the peoples of Asia. Yet this was precisely the situation which prevailed throughout the sixteenth, seventeenth, and eighteenth centuries.

The gun-carrying, ocean-going sailing ship allowed the Europeans to establish themselves as the masters of the high seas, to destroy most of the Moslem shipping and trade, and to capture a large share of the intra-Asian trade. By bringing Japanese silver to China, Japanese copper to China and India, Spice Islands' cloves to India and China, Indian cotton textiles to South-east Asia, and Persian carpets to India, the Europeans made satisfactory profits and the income derived from their shipping and commercial services paid for some of the Asian products imported into Europe. But the income was not enough and by far the greater part of Asian imports had to be paid for by a massive transfer of bullion from Europe to Asia. Large amounts of silver in the form of *reales* or pieces of eight minted in Seville, Mexican dollars, silver coins minted in Italy and Germany, French crowns and Dutch rixdollars passed every year into Asian hands. Silver was abundantly available to the Europeans, thanks to the favourable balance of trade between Europe and the Americas. Leaving aside the relatively limited trade that was carried directly via the Philippines between the Spanish colonies of the Americas and the Far East, one can justifiably say that world trade at this time consisted essentially of a large stream of silver that moved eastward from the Americas to Europe and from there to Asia and of a large flow of commodities that moved in the opposite direction: Asian products went to Europe and European products went to the Americas.

The gap in the balance of trade between Europe and Asia was conspicuous even in a physical sense. People like van Linschoten observed that when the Indiamen 'go out they are but lightly laden, onely with certaine pipes of wine and oyle, and some small quantitie of merchandie, other things they have not in, but balast and victuals for the company, for that the most and greatest ware that is commonly sent into India are rials of eight'.[2] At the end of the sixteenth century a Florentine merchant reported that Portugal and Spain alone brought into China 'every year more than

2. Van Linschoten, *Voyage to the East Indies*, vol. 1, p. 10.

1,500,000 escudos'.[3] How reliable this estimate is no one can say, but more accurate and more abundant figures are available for the seventeenth and eighteenth centuries and they consistently show that silver was the chief European export to Asia.[4]

This situation caused much concern in Europe and produced heated debates as well as numerous publications on the subject. In both England and France the companies that carried the East Indies trade were violently attacked and accused of putting their own selfish interests over and above the public interest, of thriving at the expense of national power and wealth. On a more practical level, unremitting attempts were made to improve upon the situation. In England the government required that at least one-tenth of the cargo of each Indiaman should consist of goods which were 'the growth, produce or manufacture of the Kingdome'.[5] The East India Company explored every conceivable opportunity 'to introduce [itself] into the trade of the citty of Nankin' and other cities in northern China, hoping that the cold climate would favour 'a considerable vent for our woollen manufactures'.[6] However, all these efforts ended in failure. European merchants even explored the possibility of exporting paintings and *objets d'art*, but western art was intensely concerned with religious subjects and, as Richard Cocke wrote from Japan, the Asian peoples were not interested in Biblical scenes. 'They esteem a painted sheet of paper with a horse, ship or a bird more than they do such rich picture. Neither will any one give six pence for that fair picture of the conversion of St Paul.' After having tried without success to sell engravings of Madonnas and other religious scenes, the Amsterdam Company tried to sell prints that

3. Carletti, *My Voyage around the world*, p. 153. The report by Carletti was addressed to the Archduke of Florence. Carletti obviously referred to the '*scudi d'oro*' (cf. ibid., p. 144) which were in his days the main gold currency of Florence. A Florentine '*scudo*' was worth slightly more than three grams of pure gold. Thus, according to Carletti, 'Portugal and Spain alone' brought into China every year an amount of silver roughly equivalent to about 4.5 metric tons of gold.

4. Cf. Morse, *East India Company*, vol. 1, pp. 8, 307–13 and *passim*; Chaudhuri, *East India Company*, pp. 24–5; and Dermigny, *Le Commerce à Canton*, vol. 2, pp. 687–767.

5. Morse, op. cit., vol. 1, p. 67.

6. Morse, op. cit., p. 109.

had 'a more general human appeal, a collection of nudes and less decent illustrations',[7] but these imaginative efforts also failed to achieve appreciable results. About 1701 the Council of the East India Company wrote to the court in London: 'We cannot tell what to advise your Honours to send to these parts [China], the natives being fond of nothing but silver and lead; and probably if the rest of your goods were thrown over board at sea, your cargo home would not be much less'.[8]

The lack of eastern demand for western products was a serious problem, but even more alarming was the fact that Asian manufactures competed successfully with European products in the European market in important sectors of the economy. As J. Cary, a merchant of Bristol, put it, 'the East India trade I take to be very prejudicial to us, as 'tis not driven, because it exports our bullion, spends little of our product or manufactures and brings in commodities perfectly manufactured which hinder the consumption of our own'.[9] The story of the East India silks and calicoes that were imported into England and caused difficulty for the English textile industry is so well known that it does not need to be told here. It was fortunate for England that no Indian Ricardo arose to convince the English people that, according to the law of comparative costs, it would be advantageous for them to turn into shepherds and to import from India all the textiles that were needed. Instead, England passed a series of acts designed to prevent importation of Indian textiles and some 'good results' were achieved.

Similar feelings and reactions arose in France in connexion with the importation of silk and textiles from China by the French East India Company. Interestingly enough, Colbert, the great 'mercantilist', did not assent to the petitions of the French producers who solicited protective measures, but his immediate successors did, and from 1686 onward a sequence of acts repeatedly forbade all imports of Asian textiles into France. In 1717 and 1718 also Spain passed a law prohibiting the import of silk textiles from Asia.

7. For all that precedes cf. Panikkar, *Asia and Western Dominance*, p. 53.

8. Morse, op. cit., vol. 1, p. 114.

9. Cary, *Discourse on Trade*, p. 43.

As I have said, most of the European manufactures either did not interest the peoples of Asia or could not compete with similar products of Oriental making; but there were a very few exceptions and the mechanical clock was one of them.

As Father Matteo Ricci wrote, the timepieces used in China were

made to operate by the means of water or fire, and in those which are operated by fire, time is measured by odoriferous ashes of a standard size. Besides the Chinese make also other clocks with wheels that are made to turn with sand. But all such instruments are very inaccurate. Of sundials, they have only the equinoctial ones but do not know well how to adjust them for the position (i.e. latitude) in which they are placed.[10]

Rumours of mechanical clocks reached Chinese circles from Macao. The idea of 'self-ringing bells' (自鳴鐘) intrigued the Chinese immensely and the Jesuits seized upon this interest as an opportunity for sneaking into China. As Father Ricci himself relates the story, in 1582 'a native of the Province of Fukien, named Chen Jui, was Viceroy of Kwantung and Kwangsi; he was a capable man and fond of money.' The Jesuits of Macao approached him tactfully and let him know that they wanted to present him with some gifts, including 'a clock that strikes the hours automatically, a very beautiful object such as has never been seen before in China'. Thus the Fathers went to the Viceroy and

presented him with the clock, a Venetian triangular-shaped glass which shows the delights of various colours and some other things. The Viceroy was extremely pleased with the gifts and assigned the Fathers a residence in a temple of idols called Tien nim tse, not far away from his palace. There the Fathers received visits from some of the Magistrates and other prominent citizens ... and there they lived for four or five months.[11]

They were beginning to hope that the concession might develop into a permanent grant when discouraging difficulties unexpectedly arose. For some reason the Viceroy was relieved of his office; realizing that the presence of the Jesuits in the city might not be pleasing to his successor and might actually be the object of an investigation, he requested the Fathers to withdraw. After a

10. D'Elia, *Fonti Ricciane*, vol. 1, p. 33.
11. D'Elia, op. cit., vol. 1. pp. 161–7.

month of 'great tribulations' the Jesuits were back in Macao, filled 'with great sadness'. But scarcely a week had passed since their return to the island when they were reached by a messenger of the governor of the district of Schiaochin. With the authorization of the new viceroy, the governor invited the Fathers to return to Schiaochin and to take over a piece of property granted by the state for the construction of a house.

In order that the appearance of a new religion might not arouse suspicion among the Chinese people, the Fathers did not speak openly about religious matters. They spent their time courteously receiving visitors, in studying the language of the country, the methods of writing and the customs of the people. They taught by virtue of their example what they did not preach. They wore the outer garment of the prominent persons of the country: a long robe with very ample sleeves and this impressed the Chinese people very favourably.[12]

Always willing to oblige, the Fathers took advantage of every opportunity to be as helpful and as generous as possible. In December 1583, one of them had to return to Macao to obtain financial help for the mission from the European community. The governor of Schiaochin let the Fathers know that he would dearly love to have one of the 'self-ringing bells' sold in Macao and that he was ready to pay whatever was necessary for it. The Jesuits were unable to find a clock for sale in Macao but they did find an Indian blacksmith 'of very dark complexion' who had learned from the Europeans how to make clocks. This blacksmith was brought to Schiaochin in lieu of the clock, a proof of good will which very obviously impressed the governor. Two of the best smiths in town were assigned to assist the visiting clockmaker and after a great deal of trouble a clock was eventually produced at the mission for Governor Wang Phan.[13]

Clocks were also instrumental in opening the gates of the Imperial Palace at Peking to the Jesuits. As the story goes, the Jesuits petitioned to pay a respectful visit to the imperial court and to present the emperor with two clocks and some other gifts. One of the clocks was a large iron piece, driven by weights and richly ornamented with gilded dragons, eagles and other figures; the other was a small spring-driven piece made of gilded

12. D'Elia, op. cit., vol. 1, p. 192.
13. ibid., pp. 201–12.

bronze. Both were equipped with striking mechanisms.[14]

Suspicious bureaucrats and malicious eunuchs caused all kinds of difficulties for the missionaries. At one point, in the middle of winter, Father Ricci and his companions found themselves prisoners in a fortress, and while the probability of their reaching Peking looked very tenuous indeed the probability of their being executed grew at an alarming rate. Yet at the very moment when there seemed to be no more room for reasonable hopes, events unexpectedly took a turn for the better. The Fathers were convinced that Divine Providence had come to their assistance 'in answer to the prayers of the many who were everywhere storming Heaven for the success of the expedition'. This was the age when in the West the universe was regarded as a great piece of clockwork and God as an outstanding clockmaker; perhaps this is the reason why Providence chose clocks to change the course of the events. But let Father Ricci recount the story:

One day the King of his own prompting suddenly remembered having seen a petition by some foreigners who wanted to present him with a self-ringing bell. Thus he cried, 'How is it that those foreigners do not give me the self-ringing bell?' The eunuch who always accompanies the King, replied, 'If Your Majesty does not endorse the record, how would the foreigners dare come to the Court with no licence?' Whereupon the King endorsed the record and summoned the foreigners. . . . The Fathers arrived in Peking at the end of the Chinese year, namely on the 24th day of January 1601. On their arrival they took up residence at the palace of the eunuchs, just outside the gate, and here they arranged the presents and made a list of them. The following day the presents were brought to the Imperial Palace with considerable show and excitement. . . . The big clock, however, was not set and did not strike the hours, thus the King gave orders to summon the Fathers. This was done and they came post haste. . . . The Fathers were admitted to the second wall in a courtyard where the big clock had been laid and where a multitude of people had gathered. The King sent one of his high-ranking eunuchs named Li Chin who received the Fathers very graciously. He asked the Fathers many questions and he wanted to know their purpose in bringing presents to the King. He was quite pleased to know that they were not asking for any position and that they were religious people who served God with no interest in worldly goods. He was told that the clocks served to tell the time by day and by

14. D'Elia, op. cit., vol. 2, p. 123 n.7, and p. 124 n.1.

night, either by the striking of the bells or by the hands on the dials. The Fathers also explained that the clocks had to be regulated and that in two or three days they could give anyone the necessary instructions. The eunuch reported all this to the King and the King appointed to the task four eunuchs from the College of the Mathematicians which is within the Palace and is composed of twenty or thirty eunuchs. The four eunuchs had to learn how to regulate the clocks and to bring them into the King's room within three days. Thus the Fathers were lodged for three days in the quarters of the mathematicians and they spent days and nights teaching the four mathematicians how to take care of the clocks. They were always treated with great respect and the eunuch Mathan and his group paid for everything. It had been rumoured that he had accepted many valuable presents from the Fathers, which was not true, but Mathan had to disburse much money to silence the eunuchs who are close to the King.

The four mathematicians learned very diligently everything pertinent to the clocks. They actually wrote down everything they were told for they could not afford to forget any detail since this could cost them their lives. The King is so cruel to them that even for a slight fault the poor unfortunates are beaten to death or killed in some other manner. Their first care was to ask for the names in Chinese of all wheels, pinions, keys, and every other part of the clocks, and this the Fathers did, creating names both in the spoken and in the written language. . . . The three days assigned for instruction had not passed before the King called for the clocks. They were hastily taken to him and he was so pleased with them that he promoted the four eunuchs to a higher rank, which is a way to give them more authority and more income. The eunuchs were delighted, especially because, from that day on, two of them were permitted to enter the presence of the King to wind the small clock which he always kept with him day and night . . .

There was no place in the Imperial Palace where the big clock could be set to allow the weights to fall low enough to govern the wheels. So in the following year the King sent the clock to the Department of Public Works and gave the order to build a suitable wooden tower according to the design made by the Fathers. This tower had staircases, windows, and balconies, and it is very beautiful indeed. He also had the bell of the clock replaced by a new one. The clock tower is indeed worthy of royalty and one thousand and three hundred ducats were spent on it. By royal order the clock tower was placed outside the second wall of the Palace, in one of the gardens of the King, where he often goes for recreation and where he keeps many other precious things.[15]

15. D'Elia, op. cit., vol. 2, pp. 120–28.

The passion of the Sons of Heaven for the 'self-ringing bells' never faded away. Throughout the seventeenth and eighteenth centuries clocks, automata, and similar devices of great beauty and ingenuity flowed unceasingly into the Imperial Palace in Peking. Emperor K'ang Hsi (1662–1722) went so far as to set up a workshop for the manufacture of clocks and watches in the Palace and the Jesuits, with characteristic pliancy, recruited professional clockmakers into their ranks and incorporated them into the China Mission. In 1707 an accomplished Swiss clockmaker, Father Stadlin, was dispatched to Peking to assist in the establishment and operating of the imperial workshop and until the suppression of the Jesuit Mission, there was always a Jesuit in charge of the emperor's clock collection and factory. In the 1730s Father Valentin Chalier wrote:

> The Imperial Palace is stuffed with clocks ... watches, carillons, repeaters, organs, spheres and astronomical clocks of all kind and description – there are more than four thousand pieces from the best masters of Paris and London very many of which I have had through my hands for repairs or cleaning. I must know as much of the theory now as any clockmaker in Europe, for I am sure few have had as much experience.[16]

According to Father Chalier, about one hundred Chinese worked under him during his superintendency of both the collection and the factory in the 1730s and 1740s.

Emperor K'ang Hsi was filled with admiration, rare in his dynasty, for the arts and sciences of Europe, but even the emperors who had no sympathy for the West or leanings toward Western technology never allowed their feelings to interfere with the preservation and enrichment of the fabulous imperial collection. Simon Harcourt-Smith studied the remains of the collection after its despoilment on the sack of Yuan-Ming Yen in 1860 and the Forbidden City in 1900 and its further destruction in the troubles of the early decades of the twentieth century. Even then he was able to reconstruct sufficient evidence to write that 'in the Imperial Palaces at Peking, Yuan-Ming Yen and Jehol the passage of the hours was marked by a fluttering of enamelled wings, a gushing of glass fountains and a spinning of paste stars, while

16. Pelliot, *Bulletin Critique*, p. 66.

from a thousand concealed and whirring orchestras the gavottes and minuets of London rose strangely into the Chinese air.'[17]

The interest in the 'self-ringing bells' was not confined to the upper circles of Chinese society; the populace too never failed to manifest great wonder and admiration when they viewed the new contrivance. At the Jesuits' House in Schiaochin, Father Ricci and his companions installed a public clock that struck the hours and unfailingly attracted a large number of curious idlers to the mission. Similarly in Peking the drum-operated carillons built by the Jesuits produced exhilarating effects on the populace.

There is nothing strikingly remarkable in the fact that a strange novelty should excite curiosity and wonder. However, one must remember that the Chinese were not given to admiration for foreign people and foreign products. They were actually inclined to dismiss anything which was not Chinese. True, Chinese culture had always been deeply concerned with time and astronomy and the fascination exerted by the Western mechanical clock on the cultivated Chinese, may be supposedly related to this fact. Yet, if I am not mistaken, the Chinese of the sixteenth, seventeenth and eighteenth centuries did not emphasize the connexion between the Western clock and the study of the heavens. Essentially they regarded the Western clock as a toy and only as a toy. One perceives in their attitude toward the 'self-ringing bells' the same kind of feelings that the gadgets of Archimedes and Heron aroused among the ancient Greeks and Romans.

At the end of the sixteenth century, the Florentine merchant, Carletti, noticed that the Chinese had no interest whatsoever in European products but that they

> bought lenses of all sorts, and especially coloured ones. But above all things they valued those triangular-shaped glasses which show the delights of various colours when one looks through them toward the open air in the country or elsewhere, and in which one sees the reflections of many things. And these were sold at up to 500 ducats each, and such was the wonder with which they struck those people the first time that they saw them that they began to praise them aloud, saying that what one saw by means of them was the matter from which the heavens are made.[18]

17. Harcourt-Smith, *Catalogue*, p. 1.
18. Carletti, *Voyage around the World*, p. 153.

While the Europeans were using lenses to produce microscopes, telescopes and spectacles, the Chinese delighted in using them as charming toys. They did the same with clocks. Lenses, clocks and other instruments had been developed in Europe to satisfy specific needs felt by European society in response to problems set by the European socio-cultural environment. In China the contrivances fell unexpectedly out of the blue and quite naturally the Chinese regarded them merely as amusing oddities.

Intellectuals were trained in art and philosophy, not in the sciences. As Father Ricci noticed, 'no one would labour to attain proficiency in mathematics or in medicine who had any hope of becoming prominent in the field of philosophy.'[19] Urban life did not set the tone of the national culture. In a society composed essentially of an upper class of *literati* trained in the humanities and of a large mass of peasants, who 'counted their time in days and months, not in minutes or hours', the clock had little chance to play the role of a useful, practical contrivance. For this to happen, a complete change of society, of its structure and of its needs had to occur. A machine has a practical meaning only as an expression of man's response to the problems set by his environment and by his fellow men. The Chinese could grasp the usefulness of machines pertaining to irrigation, for example, but could not understand the purpose of other Western inventions. At the end of the eighteenth century, the compilers of the *Ssu – K'u eh'üan – shu ti – yao* wrote that 'In regard to the learning of the West, the art of surveying the land is most important, followed by the art of making strange machines. Among these strange machines, those pertaining to irrigation are most useful to the common people. All the other machines are simply intricate oddities, designed for the pleasure of the senses. They fulfil no basic needs.'[20]

The Chinese economy and society did not substantially change for centuries after the arrival of the Portuguese, and in China the clock remained for centuries a curious toy. In 1769 Father Jean Mathieu de Ventavon wrote: 'I have been appointed by the Emperor as clockmaker, but I should rather say that I am here as a machinist because the Emperor expects me to produce not

19. D'Elia, *Fonti Ricciane*, vol. 1, p. 42.
20. *Ssu – K'u ch'üan-shu ti-yao*, vol. 55, Tze – pu, sect. 7, p. 5.

really clocks but curious machines and automata.'[21] Van Braam, who led a Dutch embassy to Peking at the end of the eighteenth century, wrote that one day the highest in rank among the mandarins showed him

a common square bottle which he had brought with him and in which there was a little wooden mill, turned by fine sand falling through a kind of funnel at the top of the bottle upon the ladle boards of the wheel. In short, it was one of those playthings which are to be found in a thousand different shapes and to be purchased for a trifle in a European fair. He asked me if I was acquainted with this piece of mechanism. I told him that I had seen a great number, and of a much handsomer external form. He then asked me why we had brought with us nothing of the same kind. I observed in answer that as in our country they only serve for the amusement of children, we had not supposed that they would give the least pleasure or excite the smallest attention. He assured us of the contrary, and spoke in the language of a man who thinks himself the possessor of a wonder.[22]

De Guignes, the secretary of van Braam, reported the same episode and added that

one should bring to Peking especially those playthings that European boys use to amuse themselves. Such objects will be received here with much greater interest than scientific instruments or *objets d'art*.... When Chinese people buy at great expense some mechanical pieces in Canton, they do it not in order to use them for the purpose for which they were constructed but rather to use them as toys and amuse themselves.[23]

A few years later, C. Abel observed that 'in every part of China through which the (English) Embassy passed, watches were considered as objects of the greatest curiosity. The attendants of the Embassy were perpetually requested to dispose of theirs. I was not, however, able to ascertain whether they valued them as markers of time or simply as curious baubles.'[24]

According to a long-established belief, in 1550 St Francis Xavier presented a clock to Yoshitaka Ouchi, governor of Yamaguchi;

21. Pfister, *Notices biographiques*, p. 914.
22. Van Braam, *Account of the Embassy*, vol. 2, pp. 47–8.
23. De Guignes, *Voyages à Peking*, vol. 1, p. 425.
24. Abel, *Journey in the Interior of China*, p. 82 *n*.

this is believed to be the first mechanical clock of European making to reach Japan. About forty years later there is a record of the presentation of a clock to Hideyoshi (*d.* 1598) in Kyoto by another missionary. A clock made in Madrid in 1581 was presented to Ieyasu (*d.* 1605) again by a missionary.[25]

Both in China and in Japan, the example of the missionaries was soon followed by merchants who gave precious clocks to influential people in order to obtain trading licences and commercial privileges. When embassies were sent from Europe to Asia, clocks of extraordinary craftsmanship and ingenuity were usually included among the presents to be offered to the Asian rulers. Especially in China, where the bureaucratic organization of the country offered conditions conducive to abuse and where both mandarins and eunuchs could be bribed on occasion, clocks were commonly used as gifts. Failure to present an influential person with a clock or watch could cause no little trouble to the Europeans. Father Ricci related that in 1596 one of the judges in Schiaochin 'was very angry with the Fathers because they did not give or lend to him a mechanical clock and he did not hesitate to use his authority to avenge himself . . . he summoned two of the mission's domestics to his court and had them cruelly whipped'.[26] At the end of the eighteenth century, Barrow related that the Missionaries

find it necessary to make frequent, and sometimes expensive, presents to those [eunuchs] in particular about the person of his Imperial Majesty. Should any of these gentlemen [the Missionaries] happen to carry about with him a watch, snuff-box or other trinket, which the eunuch condescends to admire, there is no alternative; the missionary takes the hint, and begs his acceptance of it, knowing very well that the only way to preserve his friendship is to share with him his property. An omission of this piece of civility has been productive of great injury to the Europeans. The gentleman who regulates and keeps in order the several pieces of clock-work in the palace assured me that the old eunuch, who was entrusted with the keys of the rooms, used to go in by night and purposely derange and break the machinery, that he might be put to the trouble and expense of repairing it. This happened to him so often that, at length, he became acquainted with the secret of applying the proper

25. Robertson, *Evolution of Clockwork*, p. 196.
26. D'Elia, *Fonti Ricciane*, vol. 1, p. 382.

preventive, which although expensive was still less vexatious than the constant reparation of the mischief done to the articles of which he had the superintendence.[27]

Until the beginning of the eighteenth century, clocks were used almost exclusively as gifts and only rarely were they the object of commercial transactions. The Court Minutes of the East India Company in London and the China Factory records give no evidence that clocks and watches were exported by the Company to China during the latter half of the seventeenth century. The English customs books of the same period show exports of timepieces to Sweden, Denmark, Germany, Holland, Flanders, Italy, Russia, Turkey, New England, the Barbados and many other places, but none at all to China or Japan.[28] The records of the Dutch East India Company show that throughout the seventeenth century only a few timepieces were sent to Japan and most of these were gifts.[29]

The fact of the matter was that not many people in China and Japan could afford the expensive 'self-ringing bells' and those who could afford them, were, especially in China, in the position of getting them as gifts. This state of affairs suddenly changed in the first decades of the eighteenth century when Europe began the production of low-priced timekeepers and the English, soon followed by the Swiss, began to export them to Canton. In the 1730s Father Duhalde wrote that the Europeans 'imported formerly [to Canton] cloths, crystals, swords, clocks, striking-watches, repeating clocks, telescopes, looking-glasses, drinking glasses, etc. But since the English come regularly there every year, all these merchandises are as cheap there as in Europe.'[30] There is

27. Barrow, *Travels in China*, p. 231.

28. The information regarding the Court Minutes and the China Factory Records were kindly supplied to me by the Assistant Keeper of the India Office Records. I owe the information regarding the English Custom books to Miss A. M. Millard, who searched for me Customs 3/1 part 2 Exports and Customs 2/5 and 2/6 at the Public Record Office in London.

29. This information was kindly given to me by Miss E. J. Karreman, who consulted for me the 'Invoices of incoming and outgoing cargoes no. 762–816' and the 'Accounts of the money expended on the Journey to Edo and on gifts for Japanese authorities 1642/3–1818' no. 1161–1305 p. 56 in the *Archive of the Dutch factory in Japan* at The Hague.

30. Duhalde, *General History of China*, vol. 2, p. 302.

exaggeration in Father Duhalde's statement but it cannot be denied that after the beginning of the eighteenth century the export of European clocks and watches, particularly to China, enjoyed a sudden and remarkable expansion.[31] By 1775 it was remarked that 'the Chinese receive now watches through Canton at good prices' and about 1805 a French observer noticed that 'watches are now sold in China at a very low price. . . . I have seen watches for more than 100,000 *livres* being sent back to London and an equal amount is still here unsold. The Swedes and the Danes have imported so many watches that these are now sold in the Chinese shops at 5 piastres the pair. Almost all these timepieces come from Switzerland.'[32]

Museums and private collections contain a large number of priceless and elaborate timekeepers that were produced in Europe for the Chinese market. These pieces, however, are a biased sample of what was actually exported to China. Extremely ornate clocks, pornographic watches, *et similia* were brought into the Heavenly Kingdom and at the end of the eighteenth century a sumptuary law tried to put an end to this trade.[33] But sources agree that the expansion of the European export trade in the eighteenth century was largely due to the sale of cheap pieces. A Swedish traveller related in the early 1750s that 'the Chinese buy large and little watches of the English. In their shops they have sometimes English watches to sell, and sometimes at moderate prices, but mostly of the worst kind.'[34] The situation lasted until the end of the eighteenth century when Barrow noticed that

> the gaudy watches of indifferent workmanship fabricated purposely for the China market and once in universal demand, are now scarcely asked for. One gentleman in the Hon. East India Company's employ took into his head that cuckoo clocks might prove a saleable article in China, and accordingly laid in a large assortment, which more than answered his most sanguine expectations. But as these wooden machines were constructed for sale only and not for use, the cuckoo clocks became all mute long before the second arrival of this gentleman with

31. Cf. Chapuis, *Montre Chinoise*, pp. 25*ff* and 51*ff*; Needham, *Heavenly Clockwork*, pp. 150–54; and Dermigny, *Le commerce à Canton*, vol. 3, pp. 1237*ff*.

32. Dermigny, *Le commerce à Canton*, vol. 3, p. 1239 and p. 1240 n. 1.

33. *Notice sur les objets de commerce à importer en Chine*, p. 270.

34. Osbeck, *Voyage to China and the East Indies*, vol. 1, p. 236.

another cargo. His clocks were now not only unsaleable, but the former purchasers threatened to return theirs upon his hands, which would certainly have been done, had not a thought entered his head, that not only pacified his former customers but procured him also other purchasers for his second cargo: he convinced them by undeniable authorities that the cuckoo was a very odd kind of a bird which sung only at certain seasons of the year, and assured them that whenever the proper time arrived, all the cuckoos they had purchased would once again tune their melodious throats.[35]

Far more timepieces were brought to China than to Japan, both as gifts and as objects of trade. This was due not only to the fact that the population of China was noticeably greater than that of Japan, but also to the fact that the Japanese had learned to make mechanical clocks on their own.

When Western clocks and Western fire-arms appeared in the Far East, the fanciful Chinese were fascinated by the clocks while the warlike Japanese were especially fascinated by the guns. The Japanese soon began the manufacture of arquebuses and apparently produced them in fair quantity. Only much later did they become interested in clocks, and then a ticklish problem presented itself. The European clock was based upon a system of twice twelve hours of equal length, whereas the Japanese reckoned time according to the 'natural' day in hours of variable length. Time from sunrise to sunset was divided into six hours, and so was the time from sunset to sunrise. The result of course was that during the summer, day-time hours were 'long' and night-time hours were 'short', while during the winter the opposite was true. Only at the time of equinoxes, were the hours of day and night of equal length. Western clocks were obviously unsuited to this system of reckoning time.

One can easily recognize three phases in the early history of the production of Japanese clocks. In the beginning, Japanese mechanics produced only a few slavish copies of European clocks without even trying to adapt the mechanisms of the clocks to the native system of time measurement. It is commonly asserted by Japanese antiquarians that the first clock made in Japan was produced toward the end of the sixteenth century by Tsudo Sukezaiema, a smith who made a faithful replica of the Western clocks

35. Barrow, *Travels in China*, p. 181.

that had been presented to Ieyasu.[36] Early in the seventeenth century, the Japanese craftsmen replaced the European hour circle and its Roman numerals by one giving the twelve Chinese signs of the zodiac and their corresponding numerals, but essentially the clocks they produced were still slavish replicas of European clockwork. A second phase opened up sometime in the course of the seventeenth century when some of the old clocks were 'converted' to a new style of motion work. The fixed hour circle with its revolving centre was removed and a revolving circle was substituted with adjustable 'hour' plates and a fixed hand. Finally, toward the end of the century, the Japanese mechanics devised the double escapement clock with one balance for the hours of the day and a second balance for the hours of the night.[37] By then the Japanese had developed a style of their own and produced clocks that were peculiarly suited to their system of reckoning time, even though they always used the Western principle of the verge and foliot device. Three types of weight-driven lantern clocks were made. The first one, hung by a silk cord or hook, was the wall-clock; the second, mounted on a truncated cone, down which the weights could hang unseen, was the pyramid type; the third placed on a table was known as the table clock. Besides these types, the Japanese originated a type of clock called 'pillar clock', because it was designed to be hung from the wooden pillars which supported the roofs on Japanese houses. Japan's paper walls were too fragile to support heavy Western-style wall clocks. The pillar clocks unlike any other clock told the time vertically. A hand fixed to the weight as it descended showed the time on a rod marked with adjustable hours. Later on, the Japanese craftsmen developed also very beautiful, pocket-sized Inro watches. These watches were made of brass, key-wound, and fitted into a traditional Japanese Inro, or pillbox case, which was attached to an adjustable cord. Since Japanese dress does not have pockets, Inros were worn around the neck or tucked into the *obi sash*.

Most of the Japanese clocks were made in Nagasaki, the gateway through which Western products and ideas continued to

36. Tsukada, *Wadokei*, p. 37; Takabashi, *Tokei Hattatzu shi*, pp. 47–8.

37. For all that precedes cf. Robertson, *Evolution of Clockwork*, pp 219–26, and 240–43.

creep into the country even after Japan was closed to foreigners in the 1620s and 1630s. Clocks were occasionally produced also in Kyoto the 'city of nobles, artisans and craftsmen', Edo (Tokyo) the 'city of *samurai*', Osaka the 'city of merchants', and the castle towns of Sendai and Nagoya. Needless to say, Japanese clock manufacturing never came close to the European development of the industry, neither for quality nor for quantity. The production of timepieces was always very small because the feudal structure of society and the unbalanced distribution of income set narrow limits to the demand for clocks. Almost all the '*Wadokei*' (the Japanese clocks) were specifically made for the local *daimyōs* and *shōguns*: numerous clocks could be found in the castles but there were almost none elsewhere. The number of artisans capable of producing clocks was always very small and those who made them were usually skilled smiths who on occasion turned out timepieces.[38] Even with all these severe limitations, the situation in Japan was strikingly different from that in China. In China, as we have seen, a factory for the production and repair of timepieces was established in the Imperial Palace at Peking at the end of the seventeenth century, but the enterprise retained a typically bureaucratic nature and it remained for a long time an isolated, centralized undertaking. In the 1750s P. Osbeck remarked that 'watchmakers are very much wanted here [in China]'.[39] In 1769 Father Jean Mathieu de Ventavon (1733–87) reported that 'the Princes and the dignitaries of the Empire call upon the Europeans to have their numerous clocks and watches repaired.... We are indeed overloaded with work.'[40] In 1775 it was noticed that the Chinese 'sell damaged watches to the Russians at very low prices because they have no craftsmen capable of repairing them.'[41] At the end of the century it was to his 'great

38. A Japanese clockmaker made all parts of a timepiece and specialization among craftsmen in regard to the production of single parts of timepieces did not develop in old Japan. There was no standardization of production and one can hardly find two old '*wadokei*' that look identical. It is generally believed that it took more than one year for a craftsman to produce a clock and it is said that some of the famous makers produced less than ten clocks during their lifetime. Cf. Tsukada, *Wadokei*, p. 15 and *passim*.

39. Osbeck, *Voyage to China and the East Indies*, vol. 1, p. 236.

40. Pfister, *Notices Biographiques*, p. 914.

41. Dermigny, *Le commerce à Canton*, vol. 3, p. 1239.

astonishment' that van Braam saw that in one town there were 'three watch-makers' shops'.[42] By then the craft of horology was taking firm root in Canton and around 1800, indulging in exaggeration, J. Barrow wrote that the Chinese 'now fabricate in Canton as well as in London and at one third of the expense all those ingenious pieces of mechanism which at one time were sent to China in such vast quantities from the repositories of Coxe and Merlin.'[43] Anyhow, in contrasting the history of Chinese and Japanese horology, one can hardly fail to notice that in China the art of clock-making developed at a much later date than in Japan; in China it failed to produce anything original, and it remained essentially limited to the area of Canton.

The reasons for the different response of the Japanese and the Chinese to Western technology are not easily analysed. In the case of fire-arms, it is obvious that the Chinese were not interested in weapons and military matters while the Japanese were. In the case of clocks, the situation was completely different. The Chinese were very interested in clocks, much more so than the Japanese. It may be argued that the Japanese took a less picturesque and more utilitarian view of the clock than did the Chinese, but the available evidence does not support this view.[44]

It has been said repeatedly that the Chinese traditionally regarded their country as the centre of the world while the Japanese were not hampered by self-centred cultural pride. The Chinese however, were fascinated by the 'self-ringing bell' and one may submit that, since they were unaccustomed to absorb foreign ideas, they could simply produce unimaginative replicas of Western specimens. The Japanese, on the other hand, had a long tradition of absorbing foreign ideas and, instead of merely copying the Western clock, were actually able to adapt it to their own needs and to create pieces of unmistakable originality.

Other reasons may be tentatively adduced to solve this cultural puzzle. I have already indicated that the Japanese used to measure time in hours of variable length. This system of 'un-

42. Van Braam, *Account of the Embassy*, vol. 2, pp. 212–13.

43. Barrow, *Travels in China*, p. 306.

44. According to Takabashi, *Tokei Hattatzu shi*, p. 52, clocks were regarded as oddities. In the eighteenth century, according to a Japanese book on horology, mechanical clocks were owned only by 'noblemen, rich merchants and collectors of curiosities' (ibid., p. 51).

equal hours' had once been in use in Europe and China as well but in both places it had been abandoned long before the sixteenth century. European clocks could easily be made to measure the time in China (although the Chinese divided the day into 12 rather than 24 equal hours) but they were not suitable to the Japanese system of reckoning time. Thus, resorting to Toynbeean terminology, one could say that the Japanese were confronted with a challenge that the Chinese ignored. The different social structures of the two countries may also be taken into consideration. In both China and Japan craftsmen were not very numerous; nevertheless, able artisans were not entirely lacking. With regard to China, Brusoni in the seventeenth century wrote that 'the Chinese craftsmen are very skilled in manual work and especially in working ivory, ebony and amber. . . . In mathematical things they are inferior to the Europeans. However, they are capable of making mechanical table clocks and they would make also small ones if they were paid as our craftsmen are.'[45] As far as I know, Brusoni never went to China and he was passing on second-hand information, but his statement does not seem unrealistic. There are no reasons to believe that human skills in China were in greater scarcity than in Japan, but there are reasons to believe that the mandarinate and the bureaucratic structure of the state hindered the unfolding of the potential capacities of Chinese craftsmen. Father Ricci noticed that 'not always the Chinese craftsmen strive to reach a perfection of workmanship. . . . This seems to be particularly noticeable when they toil for the magistrates who compel the craftsmen to work for them at a pay inferior to the usual.'[46] By the use of authority one may force craftsmen to work and to produce things according to traditional patterns, but it is much more difficult to force a man to be inventive and original. Effective demand expressed in monetary terms can be substituted in human affairs with other rewarding stimuli. But in old China this was not the case. The attitude of the mandarins toward the Chinese craftsmen cannot be simply disregarded as one of the innumerable examples of abuse of power on the part of bureaucrats. There was more to it than that. The set of socio-cultural values prevailing in Ming

45. Brusoni, *Varie osservazioni*, p. 13.
46. D'Elia, *Fonti Ricciane*, vol. 1, p. 29.

China actually penalized craftsmen and craftsmanship. As has been rightly said, 'the difference between artists and artisans was almost a difference of race' and 'a learned Chinese examining the work of an artisan will speak of it with the same astonishment as he would if he examined, say, the intelligent work of a beaver'.[47] The set of social values prevailing in Ming Ch'ing China favoured the oppression of craftsmen and obstructed the progress of applied science and technology. The relationship between the Japanese feudal lord and the craftsmen in the village below his castle was seemingly much more productive than the exacting bureaucratic relationship between the temporarily appointed mandarin and the craftsmen of the Chinese village.

To me, another and most important factor was the size of the two countries and the isolation in which most of the Chinese population lived. Macao was a Portuguese trading post and in Canton the natives were largely exposed to Western influences. But the rest of the country lived in virtual isolation. As Nieuhoff saw it, 'why these people who are ingenious and witty enough in some things are so dull and inexperienced in others, proceeds in all probability from their general averseness to deal with foreigners; it being a rule among them, to prohibit them entrance into their country, at leastwise not to admit them farther than their utmost frontiers'.[48] With the consolidation of Tokugawa rule, Japan adopted a strong policy of national isolation. In 1623 the English voluntarily left Hirado. In 1624 all Spaniards were driven from Japan and all intercourse with the Philippines was stopped. In 1636 the Japanese were forbidden to go abroad and those abroad were not allowed to return. In 1638 the Portuguese were expelled and when they sent an embassy in 1640 to re-open trade, almost the whole party was summarily executed. However, the Dutch traders were allowed to keep a trading station on the islet of Deshima in Nagasaki Harbour. Through this umbilical cord European influences continued to creep into Japan as they were creeping into China through Canton. In China, however, there were about 150 million people and internal communications with Canton were generally poor. In Japan there were less than 25 million people and the network of roads and internal

47. Prodan, *Chinese Art*, p. 26.
48. Nichuoff, *Embassy to China*, p. 166.

communications was relatively good. Despite the more intransigent and hostile policy of the Japanese government, foreign ideas and techniques could make an inroad into Japan from Nagasaki much more easily than they could into China from Canton.

All the preceding remarks are by necessity fragmentary and it is difficult to say whether or in what measure each fragment is relevant to the question of the differing responses of China and Japan to the impact of Western technology. On a more general level one may recall that Tokugawa Japan had flourishing urban centres such as Edo, Osaka and Nagasaki where a vigorous merchant class asserted and expressed itself in an ebullient urban culture that reached its apogee in the Genroku Years (1688–1704). Nothing similar occurred in China where the tone of the culture in towns was always agricultural and bureaucratic. Such considerations, however, do not answer the original question; they widen it. The fact of the matter is that the answer to the question is bound to be vague and imprecise because the question itself is vague and its legitimacy on the historical plane is rather questionable. When we ask 'why did China not *succeed* in producing clocks and guns' and 'why did China not *succeed* in moving toward industrialization' we implicitly assess Chinese performance on non-Chinese terms. But, as Collingwood wrote, we cannot justifiably assume 'that two different ways of life were attempts to do one and the same thing . . . Bach was not trying to write like Beethoven and failing; Athens was not a relatively unsuccessful attempt to produce Rome'. We should perhaps paraphrase a statement by one of the directors of the Rockefeller Foundation and conclude that 'it is not only rather graceless, but rather pointless to ask why in the course of the sixteenth, seventeenth and eighteenth centuries China did not develop European technologies. The wonder is that such things occurred at all.'

Chapter 3

Appendix on Escapements

The earliest form of motive power which appealed to the early craftsmen was the falling weight. Today this is still an ideal method as it gives an even and steady pull. How to control this force was the problem and it was solved with the invention of the verge escapement with foliot. As Alan Lloyd wrote, who invented such a mechanism 'no one knows, and no one probably ever will' but 'whatever his name, he was a perfect genius'.

Fig. 1

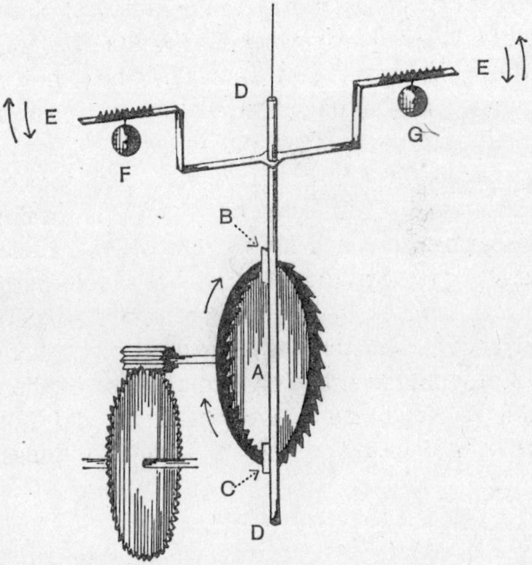

The verge and foliot device is schematically represented in Fig. 1. A wheel with an uneven number of saw-like teeth, known as the crown wheel, A, being driven round by the pull of a weight, one of its teeth comes into contact with the pallet, B, of the 'verge', DD, thus imparting circular motion to the 'foliot', EE, until the tooth escapes past the pallet, B, leaving the crown wheel, A, free to advance. Immediately, however, a tooth on the low side of the wheel, A, finds itself obstructed by the pallet, C, and the process is repeated in the opposite direction. So the 'foliot', EE, swings to and fro, while the teeth of the crown wheel escape past the pallets, B and C, one at a time. The passage of time is measured by the even beats of the balance. The position of the small weights, F and G, determines the resistance that the reciprocating balance exercises upon the crown wheel and its unidirectional movement by the alteration of the position of the weights on the foliot one regulates the speed of the clockwork. As Britten wrote: 'since the design and construction of such an escape wheel from first principles would present considerable problems, it seems not unlikely that the first attempts took the shape of a pin-wheel, the pins projecting at right angles to the

Fig. 2

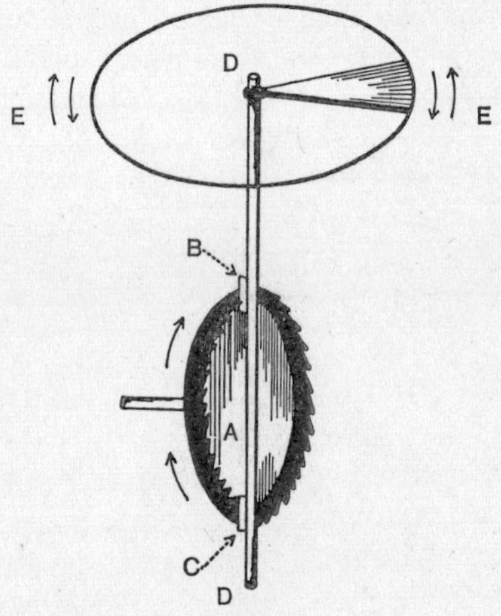

surface of the wheel. This arrangement is found in primitive Black Forest clocks quite late into the seventeenth century'.

As one can see in the manuscript by Dondi, very early the 'foliot' was sometimes replaced by a balance wheel. The balance wheel has the same reciprocating action as the foliot (see Fig. 2), but with the balance wheel regulation of the speed of the mechanism was effected by alteration of the driving weight. The question of precedence between the balance wheel and the foliot is uncertain.

Fig. 3

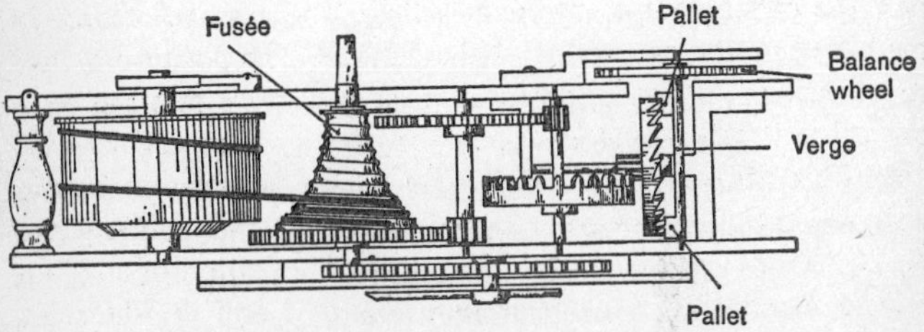

Fig. 3 shows how the verge escapement was applied to watches. The same figure also shows the *fusée*, i.e. the ingenious device that assured a uniform drive from a spring whose power decreases as it uncoils. The *fusée* is shaped like a truncated cone with spiral grooves which hold a cord or chain in place round it. This cord is joined to the barrel of the main spring. When tightly wound up

Fig. 4

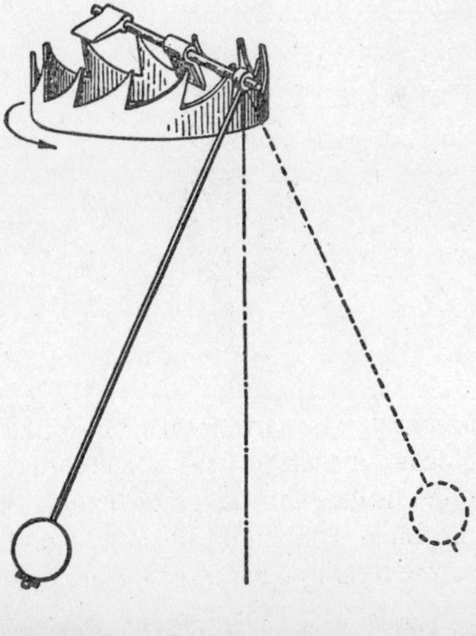

the spring turns the *fusée* by pulling the cord from the smallest diameter of the *fusée*; when it becomes wound down, it pulls the cord from the larger diameter of the *fusée*. In this way, by pre-adjusting the diameter of the *fusée* to the varying pull of each turn of the spring, the force on the main wheels of the clockwork is uniform.

Fig. 4 shows the form of the verge escapement as later adapted to the pendulum. It consists again of a crown wheel across which is positioned an arbor with two pallets set in it at right angles to each other. The motive power of a weight or of a spring causes the crown wheel to move unidirectionally, while the pendulum swings; on the other hand the swinging of the pendulum controls the movement of the clockwork and measures the time. This is the first application of the pendulum to clockwork.

Bibliography

ABEL, C., *Narrative of a Journey in the interior of China (1816–17)*, London, 1818.

ALBUQUERQUE, A., *The Commentaries* (ed. by W. de Gray Birch), London, 1875–85.

AL-QALQASHANDI AHMED IBN ALI, *Subh al-a'shā*, Cairo, 1913–19.

AMBURGER, E., *Die Familie Marselis. Studien zur russischen Wirtschaftsgeschichte* (Giessener Abhandlungen zur Agrar – und Wirtschaftsforschung des Europäischen Ostens, vol. 4), Giessen, 1957.

AMIOT, J. M., *Art militaire des Chinois* (ed. by J. De Guignes), Paris 1772 (republished in vol. 7 of the *Mémoires concernant l'histoire, les sciences, les arts, les moeurs les usages, etc., des Chinois*, Paris, 1776–91).

ATKINS, S. E., and W. H. OVERALL, *Some Account of the Worshipful Company of Clockmakers of the City of London*, London, 1881.

AUDEMARD, L., *Les jonques chinoises*, ed. by C. Nooteboom, Rotterdam, 1957.

AYALON, D., *Gunpowder and Firearms in the Mamluk Kingdom*, London, 1956.

BAASCH, E., *Holländische Wirtschaftsgeschichte*, Jena, 1927.

BABEL, A., *Histoire corporative de l'horlogerie, de l'orfèvrerie et des industries annexes*, in 'Mémoires et Documents publiés par la Société d'Histoire et d'Archéologie de Genève, vol. 33, Geneva, 1916.

BABEL, A., *Histoire économique de Genève*, Geneva, 1963.

BALLARD, G. A., *Rulers of the Indian Ocean*, London, 1927.

BANG, N. (ed.), *Tabeller over Skibsfart og Varetransport gennem øresund*, Copenhagen–Leipzig, 1906–53.

BARBOUR, V., *Capitalism in Amsterdam in the Seventeenth Century* Baltimore, 1950.

BARROW, J., *Travels in China*, London, 1804.

BASSET, M. A., Essais sur l'historique des fabrications d'armement en

France jusq'au milieu du XVIII siècle, *Mémorial de l'artillerie française*, 14 (1935), pp. 881–1280.

BEDINI, S. A., The Scent of time, in *Transactions of the American Philosophical Society* N. S. vol. 53, part 5, Philadelphia, 1963.

BELGRANO, L. T., Degli antichi orologi pubblici d'Italia con aggiunte e notizie della Posta in Genova, in *Archivio Storico Italiano*, 7 (1868), pp. 28–68.

BERGIER, J. F. and L. SOLARI, Histoire et élaboration statistique. L'exemple de la population de Genève au XVe siècle, in *Mélanges d'histoire économique et sociale en hommage au professeur Antony Babel*, Geneva, 1963, vol. 1, pp. 197–225.

BEVERIDGE, W., *Prices and Wages in England from the Twelfth to the Nineteenth Century*, London–New York–Toronto, 1939.

BOËTHIUS, B. & HECKSCHER, E. F., *Svensk Handelsstatistik 1637–1737*, Stockholm, 1938.

BOTERO, G., *Aggiunte fatte alla Ragion di Stato*, Venice, 1659.

BOTERO, G., *Relazioni Universali*, Venice, 1659.

BOURRIAU, R. P., *Notes pour servir à l'histoire des horlogers à la Rochelle du XVIe siècle au début du XVIIIe siècle*, Besançon, 1934.

BOXER, C. R., *Jan Compagnie in Japan*, The Hague, 1936.

BOXER, C. R., *Expediçoēs militares portuguêsas em auxilio dos Mings contra os Manchus 1621–1647*, Macau, 1940.

BOXER, C. R., *Fidalgos in the Far East 1550–1770, Fact and Fancy in the History of Macao*, The Hague, 1948.

BOXER, C. R. (ed.), *South China in the Sixteenth Century*, London, 1953.

BOXER, C. R., The Portuguese in the East, in *Portugal and Brazil*, ed. by H. V. Livermore, Oxford, 1953.

BOXER, C. R., *The Great Ship from Amaçon*, Lisbon, 1959.

BOXER, C. R., *Four Centuries of Portuguese Expansion, 1415–1825: a succinct survey*, Johannesburg, 1961.

BOYLE, R., *Works*, London, 1772.

BRITTEN, F. J., *Old clocks and watches and their makers* (7th edn. ed. and revised by G. H. Baillie, C. Clutton and C. A. Ilbert), New York, 1956.

BROWN, D. M., The impact of fire-arms on Japanese warfare 1543–98, *The Far Eastern Quarterly*, 7 (1948), pp. 236–53.

BRUSONI, G., *Varie osservazioni sopra le Relazioni Universali di G. Botero*, Venice, 1659.

CADDEO, R. (ed.), *Le Historie della vita e dei fatti di Cristoforo Colombo per D. Fernando Colombo*, Milan, 1930.

CAMPORI, G., Gli orologieri degli Estensi, in *Atti e Memorie delle*

RR. Deputazioni di Storia Patria per le Provincie dell'Emilia, N.S., vol. 2 (1877), pp. 243–65.

CANETTA, C., Vicende edilizie del Castello di Milano, in *Archivio Storico Lombardo*, 10 (1883), pp. 327–80.

CARLETTI, F., *My voyage around the world* (ed. H. Weinstock), New York, 1964.

CARRASCO, A., Apuntes para la historia de la fundicion de la artilleria de bronce, in *Memorial de Artilleria*, series 3, vols. 15 and 16, Madrid, 1887.

CARRASCO, A., Apuntes para la historia de la fabricacion de artilleria y proyectiles de hierro, in *Memorial de Artilleria*, series 3, vol. 19, Madrid, 1889.

CARY, J., *A discourse on trade and other matters relative to it*, London, 1745.

CHANG, T. T., *Sino Portuguese Trade from 1514 to 1644*, Leyden, 1934.

CHAPUIS, A. (with the collaboration of G. Loup and L. de Saussure), *La Montre Chinoise*, Neuchatel, 1919.

CHIANG, M., *Tides from the West*, New Haven, 1947.

CHOW WIE, *Chung Kuk Ping Ji Shü Kö*, Peking, 1957.

CHRISTENSEN, A. E., *Industriens historie i Danmark*, Copenhagen, 1943.

CLÉMENT, P. (ed.), *Lettres, instructions et mémoires de Colbert*, Paris, 1859–82.

Codex Carolinus seu volumen Epistolarum, in *Recueil des Historiens des Gaules et de la France*, vol. 5 (Paris, 1869), pp. 485–636.

COLLADO, L., *Platica manual de artilleria*, Milan, 1592.

CRAWFURD, J., *A Descriptive Dictionary of the Indian Islands and Adjacent Countries*, London, 1856.

CUNNINGHAM, W., *Alien Immigrants to England*, London, 1897.

CUNNINGHAM, W., *The Growth of English Industry and Commerce*, Cambridge, 1919.

DAHLGREN, E. W., *Louis de Geer*, Uppsala, 1923.

DANIEMEND, I. H., *Izahli Osmanli Tarihi Kronolojisi*, Istanbul, 1947.

DAUMAS, M., *Les instruments scientifiques au XVIIe et XVIIIe siècles*, Paris, 1953.

DAVARI, S., Notizie storiche intorno al pubblico orologio di Mantova, in *Atti e Memorie della R. Accademia Virgiliana di Mantova*, 1884, pp. 211–27.

DE ARTIÑANO Y DE GALDÁCANO, C., *La arquitectura naval española*, Madrid, 1920.

D'ELIA, L., *Fonti Ricciane*, Rome, 1932–9.

D'ELIA, L., *Fonti Ricciane*, Rome, 1942–9.

DE GUIGNES, C. L. J., *Voyages à Peking, Manille et l'Île de France, faits dans l'intervalle des années 1784 à 1801*, Paris, 1808.

DE LABORDE, *Notice des émaux, bijoux et objets divers exposés dans les Galéries du Musée du Louvre*, Paris, 1853.

DE LESPINASSE, R., *Les métiers et corporations de la Ville de Paris* (Coll. Histoire Générale de Paris), Paris, 1886–97.

DE MENDOZA, J. G., *The History of the Great and Mighty Kingdom of China*, ed. by G. T. Staunton, London, 1853.

DE MONTCHRÉTIEN, A., *Traicté de l'oeconomie politique*, ed. by Th. Funck-Brentano, Paris, 1889.

DE PISAN, CH., Le livre des fais et bonnes meurs du sage Roy Charles V, in *Nouvelle Collection des Mémoires pour servir à l'histoire de France*, vols. 1 and 2, Paris, 1836.

DE RADA, M., Relation of the Things of China, in *South China in the Sixteenth Century*, ed. by C. R. Boxer, London, 1953.

DE RESENDE, G., *Chronica de El-Rei d. João II*, Bibliotheca de Classicos Portuguezes, vol. 32–4, Lisbon, 1902.

DERMIGNY, L., *La Chine et l'Occident: le commerce à Canton au XVIII siècle 1719–1833*, Paris, 1964.

DE TOTT, F., *Mémoires* (II edition with the strictures of N. De Peyssonnel), London, 1786.

DEVELLE, E., *Les Horlogers Blésois au XVIe et au XVIIe siècle* (2nd edn.), Blois, 1917.

DIN TA SAN & OLESA MUÑIDO, *El poder naval chino desde su origine hasta la caida de la Dinastia Ming*, Barcelona, 1965.

DI VARTHEMA, L., *The Travels*, ed. by J. W. Jones and G. P. Badger, London, 1863.

DOORMAN, G. (ed.), *Patents for inventions in the Netherlands*, The Hague, 1942.

DOUËT-D'ARCQ, L., *Comptes de l'hôtel des Rois de France au XIVe et XVe siècles*, Paris, 1865.

DUHALDE, J. B., *The general history of China*, London, 1741.

ELIAS, J., *Schetsen uit de Geschiedenis van ons Zeewezen*, The Hague, 1916.

EVRARD, R. & A. DESCY, *Histoire de l'usine des Vennes*, Liège, 1948.

FALLET-SCHEURER, M., *Geschichte der Uhrmacherkunst in Basel 1370–1874*, Bern, 1917.

FERNANDEZ DURO, C., *Disquisiciones nauticas*, Madrid, 1876–80.

FFOULKES, CH., *The Gun-Founders of England*, Cambridge, 1937.

FILLET, Les horloges publiques dans le Sud-Est de la France, in *Bulletin Archéologique du Comité des Travaux Historiques et Scienti-*

fiques, 20 (1902), pp. 101–19.

FIRTH, B. (ed.), *Gloucestershire Marriage Allegations*, the Bristol and Gloucestershire Archaeological Society, 1954.

FRANKEL, S. H., *The Economic Impact on under-developed societies*, Oxford, 1953.

FRANKLIN, A., *La vie privée d'autrefois*, vol. 4: *La mésure du temps*, Paris, 1888.

FROISSART, J., *Li Orloge Amoureuses*, in *Oeuvres de Froissart, Poésies*, ed. A. Scheler, Bruxelles, 1870–2, vol. 1, pp. 53–86.

GARZONI, T., *La Piazza Universale di tutte le professioni del mondo*, Venice, 1595.

GEISENDORF, P. F., Métiers et conditions sociales du premier Refugé à Genève (1549–1587), in *Mélanges d'histoire économique et sociale en hommage au professeur Antony Babel*, Geneva, 1963, vol. 1, pp. 239–49.

GÉLIS, E., *L'Horlogerie ancienne*, Paris, 1949.

GENTILINI, E., *Il perfetto bombardiere*, Venice, 1626. (The first edition was published in Venice in 1592, slightly different in the form, but not in the substance.)

GILLE, B., *Les origines de la grande industrie métallurgique en France*, Paris, 1947.

GLAMANN, K., The Dutch East India Company's Trade in Japanese Copper, in *The Scandinavian Economic History Review*, 2 (1953), pp. 41–103.

GLAMANN, K., *Dutch-Asiatic Trade 1620–1740*, Copenhagen–The Hague, 1958.

GODINHO, V. M., *Les Grandes Découvertes*, Coimbra, 1953.

GODINHO, V. M., Le repli vénitien et égyptien et la route du Cap, in *Éventail de l'Histoire Vivante – Hommage à Lucien Febvre*, Paris, 1953, vol. 2, pp. 283–300.

GOODRICH, L. C., Note on a few early Chinese Bombards, in *Isis*, 35 (1944), p. 211.

GUICCIARDINI, F., Relazione di Spagna (1512–13), in *Opere* (R. Palmarocchi, ed.), Bari, 1936.

HAHR, G., *Åkers styckebruk*, Stockholm, 1959.

HALL, A. R., The Scholar and the Craftsman in the Scientific Revolution, in *Critical Problems in the History of Science* (ed. M. Clagett), Madison, Wisc., 1959.

Hankuk Haeyang Sa, (ed. by the Korean Naval Office, Heakun Benbu), Seoul, Danki 4288 (A.D. 1955).

HARCOURT-SMITH, S., *A Catalogue of Various clocks, watches, auto-*

mata and other Miscellaneous objects of European workmanship dating from the XVIIIth and the early XIXth centuries in the Palace Museum and the Wu Ying Tien, Peiping, Peiping, 1933.

HECKSCHER, E. F., *Sveriges Ekonomiska Historia*, Stockholm, 1935–49.

HEIJKENSKJÖLD, C., Svensk styckegjutning och lodstöpning av järn under perioden 1540–1840, in *Artilleri Tidskrift*, 64 (1935), pp. 57–79.

HENRARD, P., Documents pour servir à l'histoire de l'artillerie en Belgique. Les fondeurs d'artillerie, in *Annales de l'Académie d'Archéologie de Belgique*, 45 (1889), pp. 237–81.

Histoire de l'Académie Royale des Inscriptions et Belles-Lettres, Paris, 1736.

HILDEBRAND, K. G., *Fagerstabrukens Historia: Sexton-och Sjuttonhundratalen*, Uppsala, 1957.

HO PING-TI, *The ladder of success in Imperial China*, New York, 1964.

HUDSON, G. F., *Europe and China*, London, 1961.

HUMMEL, A. V., (ed.), *Eminent Chinese of the Ch'ing Period*, Washington, 1944.

HU TSUNG-HSIEN, *Ch'ou hai t'u pien* (preface dated 1624); (two copies of this work are available at the Asiatic Library of the University of California, Berkeley).

IBN KHALDUN, *Kitab al-ibar wa- Dīwān al Mubtada wal Khabr fī-ayyām al-'arab*, Cairo, 1867.

IOVIUS, P., *Historia sui temporis*, Paris, 1558.

JAKOBSSON, T., *Beväpning och Beklädnad* (vol. 2 of *Sveriges Krig 1611–1632*), Stockholm, 1938.

JANIÇON, F. M., *État présent de la République des Provinces-Unies*, The Hague, 1729.

JANSSON, S. O., *Måttordbok*, Stockholm, 1950.

JONES, R. F., *Ancients and Moderns* (Washington University Studies, N.S. Language and Literature, no. 6), St Louis, 1936.

KURTS, B. S., *Socinenie Kilburgera O Russkoj torgovle*, Kiev, 1915.

LABARTE, J. (ed.), Inventaire du mobilier de Charles V, roi de France, in *Collection de Documents inédits sur l'histoire de France*, Ser. III, Paris, 1879.

LAMBROS, S. P., Ipomnima tou Kardinaliou Vissarionos,. in *Neos Hellenomnemon*, 3 (1906), pp. 15–27.

LANE, F. C., *Venetian Ships and Shipbuilders of the Renaissance*, Baltimore, 1934.

LE COMTE, L., *Empire of China*, London, 1737.

LE GENTIL, G., *Fernão Mendes Pinto*, Paris, 1947.

LEJEUNE, J., *La formation du capitalisme moderne dans la Principauté de Liège au XVI siècle*, Liège-Paris, 1939.

LEVENSON, J. R., *Confucian China and its Modern Fate*, Berkeley, 1958.

LEWIS, M., *The Spanish Armada*, New York, 1960.

LIISBERG, B., *Urmagare og Ure i Danmark*, Copenhagen, 1908.

LLOYD, H. A., *Some outstanding clocks over seven hundred years 1250–1950*, London, 1958.

MAO YÜAN-I, *Wu-pei-chih*, 1628; (a copy of this work is preserved at the Asian Library of the University of California, Berkeley).

MAVOR, J., *An Economic History of Russia*, London–Toronto, 1925.

MAYERS, W. F., On the Introduction and use of Gunpowder and Firearms among the Chinese, in *Journal of North China Branch of the Royal Asiatic Society* (1869–70), pp. 73–104.

MICHEL, H., L'Horloge de Sapience et l'histoire de l'horlogerie, in *Physis*, 2 (1960), pp. 291–8.

MONTECUCCOLI, R., Aforismi applicati alla guerra possibile col Turco in Ungheria, in *Opere* (ed. by U. Foscolo and G. Grassi). Turin, 1852.

MONTÙ, C., *Storia della artiglieria italiana*, Rome, 1933.

MORPURGO, E., *Dizionario degli orologiai italiani*, Rome, 1950.

MORPURGO, E., *L'origine dell'orologio tascabile*, Rome, 1954.

MORPURGO, E., Alcuni appunti sugli orologiai della Volpaia, in *La Clessidra*, 15 (September 1959), pp. 23–6.

MORPURGO, E., Ruote o molle? in *La Clessidra*, 21 (September 1965), pp. 31–2.

MORSE, H. B., *The Chronicles of the East India Company Trading to China*, Cambridge (Mass.), 1926.

MORYSON, F., *Itinerary*, ed. by Ch. Hughes, London, 1903.

MOTTA, E., Musici alla Corte degli Sforza, in *Archivio Storico Lombardo*, 14 (1887), pp. 29–64; 278–340; 514–61.

MU, FU SHENG, *The Wilting of the Hundred Flowers*, New York, 1963.

MUNDY, P., *Travels in Europe and Asia*, ed. by R. C. Temple, London, 1907–9.

MURATORI, L. A., *Dissertazioni sopra le Antichità Italiane*, Milan, 1751.

MURATORI, L. A., *Rerum Italicarum Scriptores*, Milan, 1723–1770.

N.N., Tedeschi in Milano nel Quattrocento, in *Archivio Storico Lombardo*, 19 (1892), pp. 996–9.

NADVI, S. A. Z., The use of cannon in Muslim India, in *Islamic Culture*, 12 (1938), pp. 405–18.

NEEDHAM, J., *Science and Civilization in China*, Cambridge, 1954–62.

NEEDHAM, J., WANG-LING, and DE SOLLA PRICE, D. J., *Heavenly Clockwork*, Cambridge, 1960.

NIEUHOFF, J., *An Embassy to China (1655)*, London, 1669.

OAKESHOTT, M., *Political Education*, Cambridge, 1951.

OLDEWELT, W. F. H., De beroepsstructuur van de bevolking der Hollandse stemhebbende steden volgens de Kohieren van de familiegelden van 1674, 1715 en 1742, in *Economisch Historisch Jaarboek*, 24 (1950), pp. 80–161.

OSBECK, P., *A Voyage to China and the East Indies*, London, 1771.

OTTEMA, N., *Geschiedenis van de uurwerkmakerskunst in Friesland*, Assen, 1948.

P.P., L'orologio dell'Ospedale Maggiore di Milano nel sec. XV, in *Archivio Storico Lombardo*, 44 (1917), pp. 687–8.

PAGE, W. (ed.), *The Victoria History of the County of Middlesex*, London, 1911.

PANIKKAR, K. M., *Asia and Western Dominance*, London, 1961.

PARRY, J. H., *The Age of Reconnaissance*, New York, 1964.

PAUMGARTNER, H., *Welthandelsbräuche*, ed. by K. O. Müller, Stuttgart-Berlin, 1934.

PELLIOTT, P., Bulletin Critique (review of A. Chapuis, *La Montre Chinoise*), in *T'oung Pao*, ser. 2, vol. 20 (1920–21), pp. 61–8.

PELLIOT, P., Le Hōja et le Sayyid Husain de l'Histoire des Ming, in *T'oung Pao*, 38 (1948), pp. 81–292.

PENROSE, B., *Travel and Discovery in the Renaissance*, New York, 1962.

PFISTER, L., *Notices Biographiques et Bibliographiques sur les Jésuites de l'ancienne mission de Chine, 1552–1773*, Shanghai, 1932–4.

PHELPS BROWN, E. H. & S. W. HOPKINS, Seven centuries of the prices of consumables compared with builders' wage rates, in *Essays in Economic History* (ed. by M. Carus-Wilson), vol. 2, London, 1962, pp. 179–96.

PIERIS, P. E. & FITZLER, M. A. H., *Ceylon and Portugal*, Leipzig, 1927.

PIRI R'EIS, *Bahrije* (*Das Türkische Segelhandbuch für das Mittelländische Meer vom Jahre 1521* (ed. and translated by P. Kahle), Berlin-Leipzig, 1926–7.

PORTAL, R., *L'Oural au XVIII siècle*, Paris, 1950.
PRODAN, M., *Chinese Art*, New York, 1958.

RALPH, E., *Marriage Bonds for the Diocese of Bristol*, Gloucestershire, 1952.
RATHGEN, B., Feuer und Feuerwaffen des 14 Jahrhunderts in Flandern, in *Zeitschrift für Historische Waffenkunde*, 7 (1915–17), pp. 275–306.
RAYCHAUDHURI, T., *Jan Company in Coromandel 1605–1690*, The Hague, 1962.
RAYNAUD, G., Paris en 1596 vu par un Italien, in *Bulletin de la Société de l'Histoire de Paris et de l'Île-de-France*, 12 (1885), pp. 164–70.
REVERCHON, L., *Petite histoire de l'Horlogerie*, Besançon, s.d.
ROBERTSON, F. L., *The evolution of naval armament*, London, 1921.
ROBERTSON, J. D., *The Evolution of Clockwork with a special section on the Clocks of Japan*, London, 1931.
RUBBIANI, A., L'orologio del comune di Bologna e la sfera del 1451, in *Atti e Memorie della R. Deputazione di Storia Patria per le Provincie di Romagna*, ser. 3, vol. 26 (1908), pp. 349–66.

SANDOZ, CH., *Les Horloges et les maîtres horlogeurs à Besançon du XVe siècle à la Révolution Française*, Besançon, 1905.
SAVARY, J., *Dictionnaire universel du Commerce*, Copenhagen, 1761.
SCHUBERT, H. R., The first cast iron cannon made in England, in *The Journal of the Iron and Steel Institute*, 146 (1942), p. 131 P. –140 P.
SCHUBERT, H. R., The superiority of English cast-iron cannon at the close of the sixteenth century, in *Journal of the Iron and Steel Institute*, 161 (1949), pp. 85–6.
SCHUBERT, H. R., *History of the British Iron and Steel Industry*, London, 1957.
SCOVILLE, W. C., *The Persecution of Huguenots and French economic development*, 1680–1720, Berkeley–Los Angeles, 1960.
SEMEDO, A., *Histoire Universelle de la Chine*, Lyon, 1667.
SERJEANT, R. B., *The Portuguese off the South Arabian Coast*, Oxford, 1963.
SIMPSON, L. B., *The Encomienda in New Spain*, Berkeley, 1950.
SIMPSON, L. B., *Many Mexicos*, Berkeley, 1957.
SMITH, A., *The Wealth of Nations*, New York, 1937.
SMITH, J., *Old Scottish Clockmakers*, Edinburgh, 1921.
STRUMILIN, S. G., *Istoriia chernoi metallurgii v SSSR*, Moscow, 1954.
SYMONDS, R. W., *A History of English Clocks*, London–New York, 1947.

TAKABASHI, H., *Tolkei Hattatzu shi*, Tokyo, 1924.

TAWNEY, R. H. and POWER, E., (eds.), *Tudor Economic Documents*, London–New York–Toronto, 1953.

TENENTI, A., *Venezia e i corsari*, Bari, 1961.

TENENTI, A., *Cristoforo da Canal. La Marine Vénitienne avant Lépante*, Paris, 1962.

TENNENT, J. E., *Ceylon*, London, 1860.

THOMAS, P. J., *Mercartilism and the East India Trade*, London, 1963.

THORNDIKE, L., Invention of the mechanical clock about A.D. 1271, in *Speculum*, 16 (1941), pp. 242—3.

TSUKADA, T., *Wadokei*, Tokyo, 1960.

ULLYETT, K., *British Clocks and Clockmakers*, London, 1947.

UNGERER, A., *Les horloges astronomiques et monumentales les plus remarquables de l'Antiquité jusqu' à nos jours*, Strasbourg, 1931.

UZUNÇARSILI, I. H., *Osmanli Devleti Teskilatindan Kapukulu Ocaklari*, Ankara, 1943–44.

VAN BRAAM, A. E., *An authentic account of the Embassy of the Dutch East India Company*, London, 1798.

VAN DILLEN, J. G. (ed.), *Bronnen tot de Geschiedenis van het Bedrijsleven en het Gildewezen van Amsterdam*, Rijsk Geschiedekundige Publicatien, *n.*69, The Hague, 1929: *n.*78, The Hague, 1933.

VAN LINSCHOTEN, J. H., *Voyage to the East Indies*, ed. by A. C. Burnell and P. A. Tiele, London, 1885.

VASARI, *Le vite dei più eccellenti pittori, scultori, e architetti,* (ed. C. L. Ragghianti), Milan–Rome, 1942–9.

VIAL, E., and CÔTE, C., *Les Horlogers Lyonnais de 1550 à 1659*, Lyons, 1927.

VIDIER, A., Les gouverneurs de l'horloge du Palais, in *Bulletin de la Société de l'histoire de Paris et de l'Île-de-France*, 38 (1911), pp. 95–103.

Vie de Saint Louis par le Confesseur de la Reine Marguerite, in *Recueil des Historiens des Gaules et de la France*, vol. 20 (Paris, 1840), pp. 58–121.

VINCENZO MARIA DI SANTA CATERINA, *Il viaggio all'India Orientale*, Venice, 1678.

WANG-LING, On the invention and use of gunpowder and firearms in China, in *Isis*, 37 (1947), pp. 160–78.

WARD, F. A. B., *Time Measurement*, London, 1958.

WHITE, L., *Medieval Technology and Social Change*, Oxford, 1963.

WHITEWAY, R. S., *The Rise of Portuguese Power in India*, Westminster, 1899.

WILSON, C. R., *The early annals of the English in Bengal*, Calcutta, 1895–1900.

WORCESTER, G. R., *The Junks and Sampans of the Yangtze; a study in Chinese Nautical Research*, Shanghai, 1947–8.

WRIGHT, F. A. (ed. and transla.), *The Works of Liudprand of Cremona*, London, 1930.

YERNAUX, J., *La métallurgie Liègoise*, Liège, 1939.

ZINNER, E., Wurde die Räderuhr in Deutschland oder in Italien erfunden, in *Die Himmelswelt* , 53 (1943), pp. 17–22.

ZINNER, E., *Aus der Frühzeit der Räderuhr von der Gewichstuhr zur Federzugsuhr*, Deutsches Museum – Abhandlungen und Berichte 22, München, 1954.

Index

Abdul Aziz, Sultan of Turkey, 76
Abel, Clarke, English traveller and naturalist, *quoted*, 160
Abu'l-Ala al Ma'arri, Arab writer, 116
Aden, 81
Afghans, 107
Africa, Central, 104–5
Africa, North, 73, 81; *see also* Moors
Africa, West, 33, 38, 40, 42, 100, 104, 105
Agricola, Georg, German metallurgist, 19
Akers, Swedish gun factory at, 52n., 73n.
Alava, Francis de, Spanish soldier, 46n.
Al-Jazari, Moslem encyclopedist, 20
Albania, 33, 73, 104
Alberti, Leone Battista, Italian architect, 18
Albuquerque, Affonso d', Portuguese conqueror in India, 99, 102, 103, 104
Alemaigne, Jehan d', German clockmaker in France, 127
Alexandria, cannon used in (1366–68), 73n.
Alfonso d'Este, Duke of Modena, 38
Alfred, King of Wessex, 113n.
Alicante, 73
Almeida, Francisco de, Portuguese viceroy in India, 103
Al-Qalqashandi, encyclopedist, 73n.
'Aly Beg, Turkish admiral, 81
America, 104; favourable to European settlement, 105
Amiot, J. M., French Jesuit in China, 90n.
Amsterdam, main munition market of Europe, 51, 51n., 54, 54n., 57, 57n., 58, 59, 59n., 61
Amurath (Murad), Sultan of Turkey, 77
Anatolia, Turkish copper mines in, 77
Angoumois, iron ore and factories in, 61, 62
Anjou, iron industry in, 60
Antwerp, 38, 39, 40, 50, 66; decline of, 137
Archangel, 57n.
Archimedes, Greek mathematician, 15, 158
Armada, Spanish, 71, 71n.
Arnold, John, English clockmaker, 131
Artevelde, Philip van, Flemish rebel, 35

artillery, field, 38–9, 65; European superiority in, 93, 106–7
artillery, siege, 35–6, 38–9; Turkish preference for, 73–5, 76, 98; *see also* guns
Ashdown Forest, 43, 44, 45
Asser, English bishop and biographer, 113n.
Asslar, German iron works at, 51, 56
astrolabe, 120
Attila, Hun conqueror, 107
Augsburg, 39; clock industry in, 128–9, 136–7, 138
Aurangzeb, Moghul Emperor, 107
Ayalon, Dr D., *quoted*, 74
Aztecs, 105

Bach, Johann Sebastian, German musician and composer, 170
Baden, clock industry in, 145
Badoer, Venetian ambassador, 41
Baian, Avar chief, 107
Barcelona, gun foundry at, 41
Barlow, Edward, English clockmaker, 143
Barrow, John, *quoted*, 161, 163–4, 166n.
Bartolomeo di Gnudulo, Italian clockmaker, 120
Batavia, Dutch gun factory at, 86
Battles:
  Diu (1509), 80
  Keresztes (1596), 104
  Kossovo (1389), 73, 74
  Lepanto (1571), 79–80, 81
  Marignano (1515), 38
  Mohacs (1526), 74, 104
  Nicopolis (1396), 32
  Ravenna (1512), 38
  St Gothard (1664), 79
  Wahlstatt (1241), 32
Baude, Peter, English gunsmith, 45
Bayard, Master, French clockmaker in Geneva, 139
Bayazed I, Sultan of Turkey, 32
Beche, Guillaume de, gun manufacturer, 53, 61n.
Beethoven, Ludwig van, German musician and composer, 170
Beliard, *quoted*, 147
bells, connexion with guns, 36, 37, 41; role in medieval society, 114
Bergamo, 39
Berlin, clock industry in, 126, 145
Besançon, clock industry in, 127n.
Bessarion, Joanes, Cardinal (Patriarch of Constantinople), 20, *quoted*, 20–21, 24, 31, 120
Besson, Jacques, French mathematician, 19
Biringuccio, Vanuccio, Italian mathematician and metallurgist, 19
Biscay, iron industry in, 50, 61, 64
blacksmiths (as clockmakers), 125–6, 129–30, 154
Bloch Borel, J.A., 135
Blois, clock industry in, 128, 129, 131, 136, 138, 142
Bodeker, Jost, Swiss clockmaker, 125
Bohemia, copper supplies in, 37
Bohemians, 33
Bolzano, 38
bombard, large primitive gun usually made of wrought iron, 35–6, 44, 66n., 75–7, 78–9; *see also* 'Mahometta' and 'Mons Meg'
Bon, Philippe, French clockmaker in Geneva, 139
Bosnia-Herzegovina, 33, 73, 104

Botero, Giovanni, Piedmontese writer, 71–2, 77
Boulogne, 62
Bourg-en-Bresse, clockmakers in, 147
bow, as weapon, 43, 71, 86, 102
Boxer, Professor C.R., *quoted*, 149
Boyle, Robert, Irish philosopher and inventor, 27
Braam, A.E. van, Dutch traveller, *quoted*, 160, 166–7
Bränn-Ekeby, Swedish gun factory at, 64n.
brass, *see* bronze
Brenchley (Kent), gun foundry at, 49n.
Brescia, Venetian foundry at, 39
Brittany, iron industry in, 60
Britten, F. J., *quoted*, 172
bronze, used for guns, 35, 36–7ff., 40, 46, 51, 52, 61, 64, 69, 72n., 86, 94
Brouage, French cannon foundry at, 60
Browne, John, English cannon manufacturer, 46n., 47–8
Browne, Thomas, English cannon manufacturer (son of above), 48, 49n.
Brunelleschi, Filippo, Florentine architect, 18, 123, 132
Brusno, Girolamo, Venetian historian, *quoted*, 168
Bull, Randolf, English clockmaker, 141
Bürgi, Jost, Swiss clockmaker, 125
Burgundy, iron manufactures in, 61, 62
Busbecq, Ogier Ghiselin de, Flemish diplomat and traveller, 100
Byzantium Empire, superior technical achievements of, 15–16, 19–20; overtaken by the West, 20ff.; *see also* Bessarion

Cairo, cannon used in (1366–8), 73n.
Calais, siege of (1346), 36
Calicut, 87–8
Calvin, Jean, French reformer, 139
Campi, 47
candles (as timepieces), 113n., 123
cannon, first use of, 115n.; *see also* guns
cannon balls, 35, 50, 75, 77–8
cannon industry, early connexion with clockmaking, 115–16, 125–6
Canton, 84; centre of Chinese trade and clock industry, 160, 162, 163, 167, 169, 170
caravel, small light fast ship (favoured by Portuguese), 67, 68, 81n.; *see also* ships
Carletti, F., Florentine merchant, *quoted*, 151n., 158
carrack, armed merchant ship, 67 *see also* ships
Cary, J., English merchant, *quoted*, 135, 152
Castel, brothers, French clockmakers, 147
Cellini, Benvenuto, Italian engraver, 47
Ceylon, 84, 96
Chalier, Valentin, French Jesuit missionary, *quoted*, 157
Champagne, iron manufactures in, 60
Chang-Ming-kang, Chinese viceroy, 103
charcoal, used for casting iron guns, 52, 64n.; price of, 58
Charlemagne, Emperor, 19

Charles V, Emperor, 41
Charles V, King of France, 113n., 117, 122–3, 127
Charles VIII, King of France, 38
Charpentier, Pierre, French clockmaker in Geneva, 139
Chen Jui, Chinese viceroy, 153
Chiang, Dr M., *quoted*, 89, 109
China, 82–5, 88–95, 97, 101n., 103, 108–9; attitude to technical knowledge, 27–8, 168–9; early types of clock, 16, 153; attitude to foreign inventions, 158, 167; social and economic conditions hostile to adoption of Western technology, 158–9, 168–9; interest in clocks, 153ff.; tendency to regard clocks as toys, 158–60; contrast with Japan, 164, 166–70
Chü-Shih-ssü, Chinese scholar-official, 91
Cimabue (Cenni di Pepi), Florentine painter, 115
clepsydrae (water clocks), 19, 113–14, 115, 153
Clock and watch industry in:
  China, 153, 166–7; *see also* collections of clocks
  Denmark, 126, 129, 139n.
  England, 126–7, 128–9, 130, 131, 135, 140–44
  France, 116, 117–18, 119, 122ff. 127ff., 137–8, 143, 145–8
  Germany, 23, 127–8, 129, 130, 136–7, 138, 145
  Holland, 129, 130
  Italy, 21–4, 120–22, 126, 129–30
  Japan, 164–7
  Portugal, 145
  Russia, 145
  Scotland, 126
  Sweden, 127, 129, 145
  Switzerland, 126, 128–9, 130ff., 135ff., 138–40, 143ff., 163
collections of clocks:
  Chinese (Imperial), 154–8, 161, 162n., 166
Colbert, Jean Baptiste, French Minister, 59–63
Colbert, Jean Baptiste, Marquis de Seignelay, French Minister (son of above), 57n., 58, 59
Colliado, L., Spanish military writer, 39; *quoted*, 39–40
Collingwood, R., *quoted*, 170
Cologne, trade centre of German iron ordnance, 56
Colombo, 84
Columbus, Christopher, Genoese seaman, 101
compass, 66; invention of, 25
Constantinople, 31, 33, 75, 104
Copenhagen, clockmakers' guild in, 129
copper, used for bronze guns, 37, 38, 39, 40, 51, 52, 61, 101n.
Coromandel, Dutch factories at, 103
Coster, clockmaker, 143
Coudray, Julien and Guillaume, French clockmakers, 137
Cratzer, Nicholas, German clockmaker in England, 141
Cronberg, B.O., manager of Swedish cannon monopoly, 63n.
Crusades, 29–30, 100
Cruz, Gaspar de, *quoted*, 95
Cuper family, clockmakers in England, France and Germany, 142
Cuper, Lewis, French clockmaker in England, 142
Cusin, Noël, French clockmaker and gunsmith, 126

Cydones, Demetrius, Byzantine writer, 20

Dante Alighieri, Italian poet, 14, 116
Dardanelles, The, Turkish cannon at, 78
Dasypodius, Conrad, mathematician, 132
Dauphiné, iron ore in, 61
Deburges, Jacob, Anglo-French clockmaker, 131
Deccan, King of (Mahomet Shah) 83
Denmark, 22, 47, 61, 126, 129, 149, 162, 163
Diaz, Bernal, *quoted*, 99
Dinant, 39
Dondi, Giovanni de', Italian scholar, 121–2, 123, 130, 132, 172
Dondi, Jacopo, Italian scholar, 121, 132
Drondelle, Laurent, French clockmaker in Geneva, 137
Duckworth, Sir John, British admiral, 78
Duduict, Jacques, French clockmaker in Geneva, 137
Duhalde, Jean Baptiste, French Jesuit missionary and geographer, *quoted*, 162–3
Dunkerque, 62
Dürer, Albrecht, German engraver, 19
East India Company (Dutch), 151, 162
East India Company (English), 107, 151–2, 162
East India Company (French), 151, 152
Edict of Nantes, effects of Revocation of, 146
Edo (Tokyo), 166, 170
Edward III, King of England, 36n., 126
Eginhard Frankish historian, *quoted*, 19
Egypt, 80
Ehrendal, Swedish gun factory at, 64n.
Eijck, Abraham van, manager of Swedish cannon monopoly, 64n.
Elbe River, 29
Elefanten, Swedish warship, 69
Elizabeth I, Queen of England, 47, 48, 49, 51
England, 22, 37, 40; cannon industry in, 36, 42n., 43, 50, 57, 58–9, 63; ships of, 68, 69, 70–71, 102, 108; *see also* clock and watch industry, East India Company (English), London
Erghum, Ralph, Bishop of Bath and Wells, 126
Ethiopia, 86n., 102, 105
Evelyn, John, English diarist, 27
export and import of guns, 40–41, 42n., 47, 50, 51, 53–6, 58–9, 60–61

Fada, Swedish gun factory at, 73n.
Falkenberg, Conrad von, Swedish observer and diplomat, 54n.
Fatio de Duillers, Nicolas, Swiss mathematician, 133
Feng-Kuei-fen, 19th century Chinese scholar, *quoted*, 95
Ferishta, Persian historian, *quoted*, 83
Ferney, clock industry in, 145
Ferrer, Jaime, Spanish seaman, 34

Ffoulkes, Charles, Master of Armouries in Tower of London, *quoted*, 47
Filarete, Antonio, Florentine architect, 18
Finspong, Swedish gun factory at, 73n.
fire-clocks, 113, 153
Flanc, Jehan, French clockmaker, 137
Flanders, *see* Low Countries, Southern
Flensborg, guns and clocks made at, 126
Florence, 35
Fobier, Pierre de, French clock-maker in Geneva, 139
foliot (balance arm fixed across the verge at the top), 115, 116, 133, 165, 171–4
Fontana, Tommaso, Italian writer, 18
Foscari, Francesco, Doge of Venice, 29
France, 22, 23, 28, 31, 32, 33, 38, 43, 44, 45, 47, 100; cannon industry of, 59–63; ships of, 69; *see also* clock and watch industry, East India Company, (French) Huguenots, Paris.
Francesco di Giorgio Martini, Italian painter, 25; *quoted*, 18
Francis of Assisi, St, 15
Francis Xavier, St, 160
Franco-Spanish Wars (1494–1559), effect in Italy, 23
François I, King of France, 100
Frankfurt, 39
Frederick II, King of Prussia, 145
friars, as early clockmakers, 126–7, 130, 132
Friesendorff, Johan von, manager of Swedish cannon monopoly, 64n.
Froissart, Jean, French chronicler *quoted*, 117n., 118
Fromanteel, John, English clockmaker, 143
Fugger, German merchant and banking family, 38
Fuggerau, gun foundry at, 38
fusèe (mechanical device to equalize the force of the spring), 124, 144, 173–4

Galileo, Galilei, Italian astronomer, 108, 133
galleon, 69–71, 98; *see also* ships
galley, low flat single-decked vessel using sails and oars, 65–8, 70; *see also* ships
Gama, Vasco da, Portuguese explorer, 80, 99; 'era' of, 32, 108–9
Garzoni, T., Venetian writer, *quoted*, 136
Geer, Louis de, merchant and financier, 46n., 53
Geneva, centre of Swiss clockmaking, 128, 129, 130, 131, 135, 137, 138–9, 142, 143, 144, 146, 147, 148
Genoa, 31, 39, 66, 67, 101
Genroku Years (1688–1704), Japanese urban culture during, 170
Gentilini, E., Venetian gun founder, *quoted*, 45–6
Germany, 13, 23, 37, 38, 39–40, 41, 51, 56, 64; *see also* Augsburg; clock and watch industry; Nuremburg
Giotto di Bondone, Italian painter, 15, 115

Giovanni Evangelista da Piacenza, Italian clockmaker, 120
Giovio, Paolo, Italian bishop and historian, 82
Goa, 104
Goddard, John, English clockmaker, 142
Godefrin, Wathier, Flemish manufacturer, 50
goldsmiths, as clockmakers, 129, 130
governors (officials in charge of public clocks), 117, 122, 126
Grâñ, Ahmad, 86n.
Grand François, French warship, 69
Grousset, French historian, *quoted*, 32
Guericke, Otto von, German experimental philosopher, 28
Guicciardini, Francesco, Italian statesman and historian, *quoted*, 22, 38
Guidobaldo del Monte, Maria, Italian mathematician, 19
Guignes, Chrétien Louis Joseph de, French traveller and writer, 90n., 160
guilds, early organization of, 13–14; relations with clockmakers, 128–9, 131–2, 137n., 144, 146
Gujerat, 83
gunned ship, secret of European military success overseas, 68–9, 101–2, 108–9
gunpowder, invention of, 25; used by Asiatics, 82–3; used by Europeans, 101
guns:
  bronze, *see* bronze
  Chinese, 82–3, 88–93, 94–5
  Dutch, 51
  early use in Europe, 35–6; in Africa, 73; in Asia, 82–3
  English, 35–6, 42–9, 58–9
  Flemish, 35–6, 40, 49–50
  French, 38, 60–63
  German, 40
  individual, 39, 44, *see also* 'Mahometta' and 'Mons Meg'
  iron, *see* iron
  Italian, 35, 39
  Japanese, 88, 95–6
  leather, 64–5
  naval, 46n., 59, 64, 68–70, 71n., 72n., 80, 81, 83–5, 93, 94, 103
  Portuguese, 40–41, 83–4, 85n., 91
  Russian, 56–7
  Spanish, 41–3
  Swedish, 52–6, 58–9
  Turkish, 74–9, 81, 107; *see also* artillery, bombards
gunsmiths, as early clockmakers, 125–6
Gustavus I (Vasa), King of Sweden, 52
Gustavus Adolphus, King of Sweden, 53, 65

Hague, The, 51, 61; clockmakers' guild at, 129
Hall, Dr R., *quoted*, 25
Hällefors, Swedish gun factory at, 64n.
Hamburg, 61, 62n.
Harcourt Smith, S., 157–8
Haroun al-Rashid, Caliph of Baghdad, 19
Harrison, John, English carpenter and clockmaker, 131
Harry Grace à Dieu, English warship, 69

Harvey, Gabriel, English writer, *quoted*, 27
Hedda Gabler, 140
Henlein, Peter, German watch-maker, 136
Henricho, clock governor, 126
Henry II, King of France, 128n.
Henry VIII, King of England, 43, 69, 140, 141
Herlin, Chrétien, mathematician, 132
Hermuz, Portuguese expedition to, 81n.
Heron, 15, 158
Hideyoshi Tokugawa, Japanese Shogun, 161
Hirado, Dutch factory at, 103
Ho Ao, Chinese official, 84
Ho Jou, Chinese official, 88
Hocchin, Chinese armoury in, 89
Hogge, Ralph, English iron maker, 45, 47, 48
Holland, 23, 47, 48, 50–52, 53–4, 55–8, 60–61, 70, 72, 86, 88, 101, 102, 103, 106, 130, 137, 143, 149, 150, 151, 160, 162, 169; *see also* East India Company (Dutch)
Hooke, Robert, English mathematician and inventor, 133, 143
horology, mechanics of, 171–4; *see also* foliot, fusée, pendulum, spring, verge
Hsü-Kuang-chi, Chinese official, 91
Hudson, G.F., *quoted*, 106–7
Huguenots, refugee clockmakers, in Denmark, 139n.; in England, 142, 146; in Holland, 139n.; in Switzerland, 139, 146
Hundred Years War, effect in France, 22
Hungary, 37, 49, 104
Hungarians, 31, 33, 75
Huseby, Swedish gun factory at, 64n.
Huygens, Christian, Dutch physicist and inventor, 133

Ibn Buhtur, Arab chronicler, 73n.
Ibn Khaldun, Arab historian, 73n
Ibn Iyas, Arab chronicler, 73n.
Ibsen, Henrik, Norwegian dramatist, 140
Ieyashi, Japanese Shogun (founder of Tokugawa dynasty), 161, 165
India, 82–3, 85, 86–8, 95, 96, 99 *passim*, 107
Indian Ocean, 80, 102, 106
Industrial Revolution, 108
Inquisition, Spanish, 50
Ioan-Maria, Milanese renegade at Calicut, 86–8
iron, used for gunmaking:
- cast, 36, 42, 43–4, 45–7, 49–53, 54–7, 59ff., 63–4
- wrought, 36, 37, 41, 52, 53

James IV, King of Scotland, 38
Janissaries, 74–5
Japan, 51, 83, 85, 86, 88, 95–6; first European clocks in, 160–61; preference for guns, 164; manufacture of own clocks, 164–7; contrast with China, 166–70; expulsion of Europeans from, 169
Java, 83, 85
Jehan of Paris, French clock-maker, 124
Jehol, imperial palace at, 157
Jenghiz, Khan, Mongol chief, 107
Jesuits, missionaries in China, 88–90, 92, 93, 95, 96, 153–60, 161ff.; in Japan, 160ff.

Jews, clockmakers in Catalonia, 126
John I, King of Aragon, 119
John II, King of Portugal, 38, 68
Julita, Swedish gun factory at, 73n.
Junk, Chinese, 93–5; Indian, 81n.

Kang Hsi, Chinese emperor, 157
Khidr Beg, Turkish admiral, 81
Korea, 83, 86
Kuei-lin, defence of (1648), 91
Kupavno (Russia), clock factory at, 145
Kyoto, 161, 166

La Rochelle, clockmaking industry in, 137
Le Comte, Louis, French Jesuit traveller and missionary, *quoted*, 92
Le Havre, cannon manufactures at, 60
Leibnitz, Gottfried Wilhelm, German mathematician, 133
Lendenaria, John de, Italian clock governor, 117n.
Leonardo da Vinci, Italian painter and inventor, 18, 19, 25, 132
Le Roy, Julien, French clockmaker, 147
Leti, Gregorio, Italian historian, 140
Levett, William (Rector of Buxted), sub-tenant of iron works, 44
Li-Chin, Chinese eunuch, 155
Liége, 49, 50, 56
Lietuyt, John, Dutch clockmaker in England, 126
Linschoten, J.H. van, Dutch traveller, *quoted*, 95, 150
Lin tse-hsü, Chinese official, 108
Lisbon, 126, 145
Liudprand of Cremona, Lombard bishop and historian, *quoted*, 19–20
Lloyd, A.H., *quoted*, 122, 171
locksmiths (as clockmakers), 125, 126, 127, 129–30, 131
London, centre of English clockmaking, 128, 129, 137n., 139, 140–42, 143ff., 162, 163
Lopes de Sequeira, Diego, Portuguese admiral, 81n.
Lorraine, iron manufactures in, 60
Lothar, King of France, 113n.
Louis IX, King of France, 113n.
Louis XI, King of France, 124
Louis XIV, King of France, 60
Louis II, King of Hungary, 104
Louvre School, 47
Low Countries, Dutch, *see* Holland; Southern (Spanish), 39–41, 42, 44, 49, 50
Lübeck, guns imported from, 54n., 62–3n.
Lyndsay, Alexander, Scottish friar and clockmaker, 127
Lyons, 38; clockmaking industry in, 118, 127n., 128, 129, 130–31, 137n., 138, 139
Lyonnais, iron manufactures in, 61

Macao (Portuguese colony), 88, 91, 94, 103, 153–4, 169
Madrid, clock made in, 161
Maestricht, 51
'Mahometta', Turkish gun, 75, 78
Maizières, Philippe de, French writer, *quoted*, 121–2

Malabar, 88
Malacca, 84, 99
Male, Louis de, Flemish commander, 65
Malines, Spanish gun foundry at, 39, 44, 49
Malta, Knights of, 67, 81
Mamluk, 73n., 97n.
Mamluks, 73, 74, 75, 82, 97
Manchester, 108
Manfredi, Bartolomeo, Italian clockmaker, 118
Manoel I, King of Portugal, 99, 100
Marathas, 104, 107
Mariano, I., 18
Marienburg, 39
Mariotte, Jean, iron manufacturer at Liége, 56
Marsberg (Westphalia), Dutch iron works at, 51, 56
Martinot, Bernaby, French clockmaker in England, 142
Mathan, Chinese eunuch, 156
mathematics, connexion with clockmaking, 26ff., 132ff., 156, 159
Maximilian, Emperor, 38
Mazarin, Cardinal Giulio, French statesman, 60
Medici, Lorenzo I (the Magnificent), 132
Medina del Campo, Spanish gun foundry at, 41
Merfelden, Gert, Dutch gunsmith and clockmaker, 126
Mexico, 105
Milan, 39, 86, 116, 118, 127
Millimete manuscript, 35
mills, role in medieval society, 16
Mir Hussain, Egyptian admiral, 80
missionaries, in Asia, 149ff.; *see also* Jesuits
Moghuls, 96, 104
Mohammed II, Sultan of Turkey, 75, 78
Mohammed IV, King of Granada, 73
Mongols, 32, 91
Mons, 39
'Mons Meg' (English gun), 36
Montecuccoli, Raimondo, Imperial general, *quoted*, 79
Moors, 73, 81
Moryson, Fynes, English traveller, *quoted*, 24, 137, 140
Moslems, *see* Turks
Mu Fu-Sheng, Chinese writer, *quoted*, 51
Mudge, Thomas, English clockmaker and inventor, 131
Muscat, 81

Nagasaki, Dutch factory at, 86; chief Japanese centre of clock industry, 169–70
Nagoya, 166
Namur, 39
Nanking, 151
Naples, 39
Nåvekvarn, Swedish gun factory at, 64n.
navigation, difference between Atlantic (Northern) and Mediterranean, 66–8, 70, 101–2
Nawe, Francois, English clockmaker, 141
Needham, Professor J., *quoted*, 27, 116
Negroponte, the, 31, 104
Nelson, Horatio, Earl, British admiral, 66
Netherlands, *see* Holland
Newbridge, iron works at, 44

Newsam, Bartholomew, English clockmaker, 141
Newton, Sir Isaac, English astronomer, 108
Nieuhoff, Johannes, Dutch traveller, quoted, 169
Ning-yüan, defence of, 93
Nieuwenroode, Cornelis, Dutch factory manager at Hirado, *quoted*, 103
Nivernais. iron ore and gun factories in, 61, 62, 63
Normandy, iron industry in, 60
Nuremberg, clock industry in, 38, 39, 136, 137, 138
Nuyts, Pieter, Dutch Governor of Taiwan, 86
Nyenskans, 55
Nyköping, 61n.

Ögödai, Mongol chief, 32
Opium War, 108
Orban, Hungarian gunsmith in Turkish employ, 75–6
Orihuela, 73
Osaka, 166, 170
Osbeck, P., English traveller, *quoted*, 166
Ottoman, military supremacy over Mamluks, 73, 74, 77, 82; *see also* Turks
Oudenarde, siege of (1382), 35
Overtwater, Pieter Antoniszoon, Dutch factory chief at Nagasaki, 86

Paleologi, family reigning in Constantinople (1261–1453), 75
Paleologos, Constantine, Despot of Morea, 20
Palladio, Antonio, Italian architect, 18
Paris, centre of French clock industry, 117, 118, 124, 127, 128, 129, 130, 136, 138, 139, 142, 145–8
Paul I, Pope, 113n.
Peake, Edward, 48n.
Peking, 88, 91, 154ff., 160, 166
pendulum, first use of, 132–3, 143, 174
Pepin, King of the Franks, 113n.
Peres, Fernaõ, Portuguese admiral, 84
Périgord, iron ore and gun factories in, 61, 62, 63, 64
Persia, 85–6n.
Persian Gulf, 80, 81
Peter the Great, Tsar of Russia, 106
Petrarch, Francesco, Italian poet, 35
Philip IV, King of France, 122–3
Philippines, 94
phosphorus, presence in iron ore, 62
Piero Antonio, Milanese renegade at Calicut, 86–8
Pinto, Fernaõ Mendes, Portuguese writer, *quoted*, 95
Piri, 81
Pisan, Christine de, French writer, *quoted*, 123
Pius II, Pope, 33
Pius V, Pope, 79–80
Plato, Greek philosopher
Pohlem, Christopher, Swedish engineer and inventor, 145
Polo, Marco, Venetian traveller, 20, 115
Poppenruyter, Hans, Flemish gunfounder, 44
population (16th century), of America, 99–100; of Europe, 32, 43

Portugal, 31, 38, 40, 41, 54n., 68, 69, 72n., 80, 81, 145, 149, 151n., 169
Powers, H., *quoted*, 26
price of:
  bronze, 39
  cannon, 57, 58–9
printing, invention of, 18, 25n., 27–8
Public Clocks – List of those referred to in text:
  Aberdeen (town), 127
  Autun (cathedral), 126
  Beauvais (cathedral), 116
  Berne, 120
  Bologna, 117, 120
  Castronovo, 117n.
  Chartres (cathedral), 117
  Cluny (monastery), 117
  Dijon, 119
  Dundee (church), 126
  Ferrara, 117
  Geneva (church), 139
  Genoa, 117
  Ghent (belfry), 121
  La Rochelle (town), 137
  Lübeck, 120
  Lund, 120
  Lyons, 118, 127n.
  Mantua, 118
  Milan (church), 116, 118, 126
  Montelimar, 119
  Nonsuch (palace), 141
  Orvieto, 120
  Padua, 117
  Paris (bridge), 118; (palace), 118, 119, 127
  Peking (palace), 156
  Perpignan (castle), 119
  Reggio, 120
  Romans, 118–19
  Rouen, 119
  Salisbury (cathedral), 126
  Schiaochin (mission), 158
  Strasbourg (cathedral), 120, 132
  Uppsala (cathedral), 127
  Vincennes (castle), 117
  Wells (cathedral), 119n., 120, 126
Pythagoras, Greek philosopher, 120

Quare, Daniel, English clockmaker, 143

Rada, Martin de, 89
Raleigh, Sir Walter, English seaman, 48
Ramsay, Patrick, Scottish clockmaker, 126
Randers (Denmark), clock factory at, 126
Red Sea, 80, 81
Reformation, effect of on clockmaking, 138–9; *see also* Huguenots
Regent, English warship, 68
Religion and colonial expansion, 100–101; *see also* Jesuits
Requesens, Luis de, Spanish Governor of Netherlands, 49
Resende, Garcia de, Portuguese historian, *quoted*, 68
Reval, 55
Rho, Giacomo da, Italian Jesuit, 88
Ricardo, David, English economist, 152
Ricci, Matteo, Italian Jesuit missionary, *quoted*, 89, 93, 153, 154, 155–6, 158, 169, 161, 168
Richard of Wallingford, English friar and clockmaker, 127
Richelieu, Armand du Plessis, Cardinal-Duke of, French statesman, 60

Riga, 55
Roman, Geronimo, factor of Philippines at Macao, 94
Romano, Guilio, 47
Rome, 127
Rostock, 55
Rotterdam, gun foundry at, 51
Rouen, clockmakers in, 119
Russia, 64, 77, 106, 107

sand glasses, 113, 160
Sao João, Portuguese warship, 69
Savary de Brulons, Jean, French economist, *quoted*, 145
Saxony, 37
Schaedel, Dutch gunner, 88
Schall von Bell, Adam, German Jesuit, 90n.
Scientific Revolution, 25–8, 132ff.
Scotland, *see* clock and watch industry
Scoville, Professor W.C., 146
Schiaochin, Jesuit mission at, 154, 158
Semedo, A., Jesuit, 93
Sendai, 166
Serbia, 33, 73, 104
Seville, gun foundry at, 42
shadow clock (Egyptian), 113n.
ships and shipbuilding, 65–72
  Chinese, *see* Junks
  Dutch, 70, 72
  English, 67–9
  French, 66n., 69
  Genoese, 66, 67
  Japanese, *see* Junks
  Portuguese, 68, 69, 72n., 80–81
  Spanish, 71, 72, 80
  Swedish, 69
  Turkish, 67, 80, 81
  Venetian, 66, 67, 70, 75
Shoguns, Japanese rulers of Tokugawa clan, 104
Sīdī' Alī Ra'īs, 81n.
Sigtersson, Anders, Swedish workman, 62n.
Smith, Adam, British economist, *quoted*, 10ɓ, 144
smiths, *see* blacksmiths, goldsmiths, gunsmiths and locksmiths
Soares de Albergaria, Lopo, 81n.
Somellier, Master, French clockmaker in Geneva, 139
Southampton, 66n.
Sovereign, English warship, 68
Spain, 33, 37, 40, 41–3, 44, 47, 49–50, 70–71, 72, 100
specialization in watch manufacture in Europe, 143–4
Spice Islands, 31, 32, 38
spring (making portable clocks possible), first use of, 123–5
spring makers, 144
Stavsjo, Swedish gun foundry at, 64n.
Stockholm, 35, 65; clockmakers' guild at, 129; gun foundry at, 65
Stromberg, iron works at, 56
Suleiman, Sultan of Turkey, 104
Sully, Henry, English clockmaker in France, 147
sulphur, 62
Sun-Yuän-hua, Chinese official, 91
Sussex, iron industry in, 43–5
Svarta, Swedish gun factory at, 64n.
Sweden, 22, 23, 24, 47, 51, 52–6, 57, 58–9, 61, 61n., 62, 63, 64, 65, 120, 127, 145, 149, 162, 163; *see also* clock and watch industry
Switzerland, 23; *see also* clock and watch industry

Taiwan, 86
Tartars, 106–7
technological progress in the Middle Ages, 15–19, 20ff.
Thirty Years' War, effects in France, 60; Germany, 56, 137; Sweden, 53, 56
Thomson, Richard, 42n.
time-measurement, difference between European and Chinese, 167–8; and Japanese, 164–5, 167–8
Timur Leuk (Tamerlane), Mongol chief, 30, 107
tin, 37, 39, 40, 52, 61
Tokyo, 166; *see also* Edo
Tompion, Thomas, English clockmaker, 131
Tott, François, Baron de, French diplomat and traveller, *quoted*, 77–8, 79, 107
Toulouse, clockmakers in, 129
Tournai, 39, 65
Tower of London, 44, 47, 76
towns, growth of in Middle Ages, 12ff., 115
Trip, Elias, Dutch merchant, 53–4
Trip, Jacob, Dutch merchant, 53
Tsudo Sukezaema, Japanese smith, 164
Turkey, 31, 67, 73–82, 83, 95, 97–8, 104, 106, 107, 148, 162
Turkish-Russian War (1768–74), 77–8
Tyrol, 37, 38
Ubaldinus de Florentia, Italian gunsmith and watchmaker, 126
Uffelen, Abraham van, Dutch colonial Governor, 103
United Provinces, *see* Holland
Urban II, Pope, 30
Urseau, Nicholas, clockmaker to Elizabeth I, 141
Varthema, Ludovico di, Italian traveller and writer, *quoted*, 86–8
Vasari, Giorgio, Italian painter and biographer, 132
Vendramin, Venetian ambassador 43
Venice, 3, 39, 40, 66, 67, 70, 75, 80–81, 88, 98, 104; clockmakers in, 127
Ventavon, Jean Matheiu, French Jesuit and geographer, *quoted*, 159–60, 166
Verbiest, Ferdinand, Flemish Jesuit missionary, 89, 90
verge (vertical spindle fixed diametrically across the crown wheel), 115–16, 125, 165, 171–3
Vic, Henry de, German clockmaker in France, 127
Victoria, Queen of England, 76, 108
Vienna, sieges of 1529 and 1683, 104
Vijayanagar, Rajah of, 83
Vikings, 29
Villard de Honnecourt, 20
Villayer, Monsieur de, 143
Vincenzo Maria di Santa Caterina, Carmelite missionary, 72n.
Virtú, Count of, *see* Visconti, Giangaleazzo, Duke of Milan, Prince of Pavia and Count of Virtú, 121
Vivaldi, brothers, explorers, 32
Vives, Juan Luis, Spanish humanist, 19
Voltaire, François Marie Arouet de, French writer, 76, 145
Volturio da Rimini, Italian writer, 19

Vueman, John and William, Dutch clockmakers in England, 126

wadokei (Japanese clocks), 165–6
Walloons, immigrants into Sweden, *see* Beche and Geer
Wang-Hong, Chinese scholar and official, 84–5, 91
Wang Phan, Chinese governor, 154
watches (pocket), 124ff., 129–30, 135–6, 141, 143ff., 157, 162–3, 165
water clocks, *see* clepsydrae
weights, use of in clocks, 122ff., 154, 165, 171–2, 173, 174
Weinhär, iron works at, 56
Wendelin, Godefroy, 133
White, Professor L., *quoted*, 122
William of Malmesbury, English chronicler, 32
Wismar, 55
Worth, Forest of, 45

Yedo (Tokyo), 88; *see also* Edo
Yolens, Jacques, French gunsmith and clockmaker, 125
Toshitaka Ouchi, Chinese governor, 160
Yuan Ch'ung-huan, 93
Yuan-Ming Yen, imperial palace at, 157
Yung-Loh, Emperor of China, 90

Zaanstreek, 139n.

More About Penguins and Pelicans

Penguinews, which appears every month, contains details of all the new books issued by Penguins as they are published. From time to time it is supplemented by *Penguins in Print*, which is a complete list of all books published by Penguins which are in print. (There are well over three thousand of these.)

A specimen copy of *Penguinews* will be sent to you free on request, and you can become a subscriber for the price of the postage – 4s. for a year's issues (including the complete lists). Just write to Dept EP, Penguin Books Ltd, Harmondsworth, Middlesex, enclosing a cheque or postal order, and your name will be added to the mailing list.

Some other books published by Penguins are described on the following pages.

Note: *Penguinews* and *Penguins in Print* are not available in the U.S.A. or Canada

Religion and the Rise of Capitalism

R. H. Tawney

Religion and the Rise of Capitalism is a study of religious thought on social issues during the three centuries from the later middle ages to the early eighteenth century. Starting with an account of medieval theories of social ethics, it goes on to examine the impact on traditional doctrines of the new forces released by the economic and political changes of the age of the Reformation. The social backgrounds and teaching of Luther, Calvin, and the English divines from Latimer to Laud, receive attention in turn. A chapter on the Puritan Movement discusses, among other topics, the theory that Capitalism had Puritanism as one of its parents.

Turning to the practical realities of history, Professor Tawney explains the conditions which gave point to prohibitions of usury and to the insistence on a just price: describes the social consequences of the Tudor land question; and touches on the impetus to economic speculation given by the price-revolution, the expansion of foreign commerce, and the growth of the money-market.

The Western Intellectual Tradition

J. Bronowski and Bruce Mazlish

'The steam engine helped to shape the modern world at least as much as Napoleon and Adam Smith' – Sir Charles Snow

In this study of the development of ideas from the Renaissance to the opening of the nineteenth century Dr Bronowski, who is as well known as C. P. Snow as an integrator of science and the humanities, has collaborated with a distinguished American historian and philosopher. They have tried to see all history, and certainly all intellectual history, as a unity, with religion influencing government, science affecting religion, technology shaping society, philosophy changing the arts, and each playing on all the others in a complex and evolving pattern in which the mind of man has been supreme. This volume realizes their aim with impressive thoroughness and success, and it is notable for its excellent accounts of the lives and thoughts of the most influential European minds – from Leonardo to Galileo, Cromwell to Rousseau, and Adam Smith to Hegel.

'A splendid book.... Their method is that of the intellectual biography ... the emphasis is on inter-relationship, especially that of science and the humanities' – *Guardian*

'Written with clarity, modesty, and force, all rare qualities in such literature' – *Sunday Times*

'A history of ideas with a rare comprehensiveness – *Observer*

Not for sale in the U.S.A. or Canada

Also in Pelicans by Carlo M. Cipolla

Literacy and Development in the West

Mass literacy is scarcely a century old. How did this skill, which had always been a bulwark of civilized values, become also an important agent of social progress?

Professor Cipolla approaches this question with the quantitative method which he used in his earlier Pelican, *The Economic History of World Population.* He ranges from medieval monastic communities, where literacy was the sacred monopoly of small elites, through the bustle of nascent capitalism and the polemics and printing presses of the Reformation, to that great watershed in popular education, the Industrial Revolution.

This book is an object-lesson in the craft of the historian. From European recruitment papers, statistics, marriage registers and legal documents the author has isolated and reconstructed a crucial but neglected process of historical change – the transformation of a world bound by custom, ceremony and spectacle into a sophisticated technical civilization under the rule – and sometimes the tyranny – of the printed word.

Also in Pelicans by Carlo M. Cipolla

The Economic History of World Population

This book presents a global view of the demographic and economic development of mankind. Professor Cipolla has deliberately adopted a new point of view and has tried to trace the history of the great trends in population and wealth which have affected mankind as a whole. For it would have been inadequate to regard such a global history as being merely the sum total of national economic histories in abridged form. Among the massive problems that face the human race the author emphasizes the demographic explosion, the economic backwardness of vast areas, the spread of industrial revolution and of technical knowledge. Whilst the theoretical approach can help our analysis of these problems, Professor Cipolla believes that they can only be wholly grasped and solved when they are studied in their full historical perspective.